Introduction to
Cell
Biology

Introduction to
Cell
Biology

John K Young
Howard University, USA

World Scientific

NEW JERSEY · LONDON · SINGAPORE · BEIJING · SHANGHAI · HONG KONG · TAIPEI · CHENNAI

Published by

World Scientific Publishing Co. Pte. Ltd.

5 Toh Tuck Link, Singapore 596224

USA office: 27 Warren Street, Suite 401-402, Hackensack, NJ 07601

UK office: 57 Shelton Street, Covent Garden, London WC2H 9HE

British Library Cataloguing-in-Publication Data
A catalogue record for this book is available from the British Library.

INTRODUCTION TO CELL BIOLOGY

ISBN-13 978-981-4307-31-4
ISBN-10 981-4307-31-9
ISBN-13 978-981-4307-32-1 (pbk)
ISBN-10 981-4307-32-7 (pbk)

Typeset by Stallion Press
Email: enquiries@stallionpress.com

Printed by FuIsland Offset Printing (S) Pte Ltd. Singapore

ACKNOWLEDGMENTS

I would like to dedicate this book to my life companions, my wonderful wife Paula and my two fine sons, Michael and Matthew. I also much acknowledge the influence of the inquisitive minds of my students over these many years and the guiding influence of my colleagues in the Department of Anatomy.

CONTENTS

PREFACE

The first decade of the twenty-first century has witnessed an explosion of information about biology in general and about cells and tissues in particular. While this might arguably make the current time the most exciting of all for biologists, it paradoxically also makes it more difficult to teach biology. How can a teacher decide what to include and what to discard when writing a text or a lecture? What is the most essential information for an audience, and how can a student be kept alert and interested while being bombarded with an avalanche of facts?

My own solution to this problem, both for myself and my students, has been to focus on questions more than answers. Many texts put so much emphasis on what we know that they leave the impression there is little more to learn, except for some fine tuning of our knowledge. This is far from true, and I hope to demonstrate this in this abbreviated text of cell biology. My goal in writing this text is to keep alive the sense of wonder and puzzlement that naturally comes to any observer of living things. The quest for solutions to riddles of nature is often more exciting than finding the answers themselves. That is how I motivate myself to continue to understand cells, and I think this is a good way for a student to find continued motivation to learn.

This book assumes that the reader has a basic understanding of the components of cells — the nucleus, DNA, and organelles like mitochondria, the rough ER, and the Golgi apparatus. When writing this book, I wanted to avoid repeating the basic information found in introductory texts of biology. This book also focuses upon what I know and enjoy the most: the incredible variety of cell types found in the human body. I don't propose to even try to review the vast amounts of information about the molecular biology of micro-organisms or plant cells or about the biochemical pathways utilized in cells. Also, I haven't provided structural details about the molecules composing the cell membrane or cellular proteins, which can best be found in a text of structural biology. My main purpose is to try to bring alive the basic facts presented in courses in histology or cell biology by putting them in the context of challenging questions about cells.

In many courses, the varieties of cell shapes and sizes are presented in rather a matter-of-fact way. However, if human beings commonly underwent similar changes, and grew to heights of fifteen feet or acquired multiple heads or became filled with a thousand pounds of fat, the reasons for these changes would be of intense interest. I feel that the reasons why cells enlarge or acquire multiple nuclei are just as compelling and hope that these questions form a successful basis for this book.

The reader may find some of the statements in this book to be inadequately detailed, confusing, or simply hard to believe. To offset these difficulties, I have tried to provide numbered references to the scientific literature within the text of each topic. These scientific articles may be easily accessed on a website provided by the National Library of Medicine called www.pubmed.gov. Simply type in the name of the author and the date, and an abstract of the article (or often, the article in entirety) pops up.

Some of the opinions I express in this book about cells may prove to be wrong. However, I feel — perhaps erroneously! — that having a wrong opinion about a controversy is preferable to having no opinion at all. A wrong opinion mainly connotes a misplaced enthusiasm for a problem, which can be tested experimentally. A lack of an opinion connotes a lack of interest, which leads nowhere.

Chapter 1

HOW DO CELLS REGULATE THE POSITIONS AND AMOUNTS OF THEIR ORGANELLES?

All cells in the body are descended from undifferentiated cells of the early embryo called embryonic stem cells. These cells have a relatively simple, spherical shape and a balanced collection of all the organelles necessary to maintain a simple life. This is not to say that even the organelles of a simple cell are not impressive; indeed, they are enormously complex and highly organized. But a key to further differentiation of all cells is a reworking of the amounts and positions of organelles so that the cell can take on a specialized function. Each cell type has its own characteristic complement and ratio of organelles, in addition to its own shape and size of the nucleus. How is this accomplished? Only partial answers to this question are known at the moment.

1. Regulation of Nuclear Shape and Function

While the nuclei of all cells (except for germ cells) contain exactly the same number of chromosomes, the shapes of nuclei vary greatly between cells, and the functions of nuclear genes change tremendously, depending upon which cell type is being examined. For example, nuclei of muscle cells are characteristically oval-shaped and pale-staining (see Fig. 3.5), nuclei of neurons are large, round, and pale-staining and have prominent nucleoli (see Fig. 6.1), and nuclei of plasma cells are rounded and dark-staining (see Fig. 1.7). These distinctive nuclear characteristics are the most important features that allow cell biologists to distinguish between one cell type and another when looking through the microscope.

1

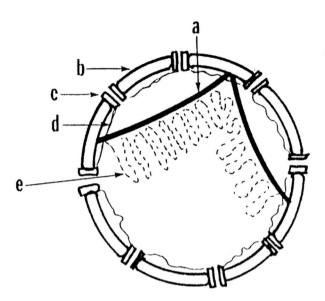

Fig. 1.1. A simplified diagram of a cell nucleus, showing (a) proteins of the nuclear matrix, (b) the outer nuclear membrane of the nuclear envelope, (c) a nuclear pore, (d) lamin proteins, and (e) chromatin.

What is responsible for this re-shaping and re-tasking of nuclei that occurs as cells differentiate? To answer this question, we must start with a review of the basic structure of the nucleus (Fig. 1.1).

The first structure we need to discuss is the nuclear envelope, which is peculiar because it consists of *two* unit membranes separated from each other by a space, the perinuclear cisterna. Most cellular organelles are bounded by only a single unit membrane; the one other exception to this rule is the mitochondrion.

The explanation for the two membranes of the mitochondrion is now well established: mitochondria arose in antiquity when a primitive cell engulfed bacteria, which survived to become organelles. The supporting evidence for this conclusion is pretty good:

- mitochondria possess a small loop of DNA that resembles the DNA loops of bacteria, and which codes for 13 of the 80 mitochondrial proteins that regulate electron transport during ATP synthesis.

- to produce energy, mitochondria expel hydrogen ions from their interior to the intermembrane space, and then allow them to trickle back into the interior via pores (ATP synthase molecules) that use the energy in this flow of ions to generate ATP. Many bacteria have analogous hydrogen-ion export systems.[3,16]

If this explanation applies to double-membraned mitochondria, does it also apply to the nucleus? Some investigators have speculated that the nucleus arose when a swarm of eubacteria engulfed another ancient organism called an archaebacterium.[20] Thus, the outer nuclear membrane represents the cell membrane of the surrounding organism and the inner nuclear membrane represents the membrane of the ancient, engulfed archaebacterium. The acquisition of definitive evidence for or against this theory has proven to be more difficult than for the theory of the origin of mitochondria.

The next nuclear component to be considered is the nuclear pore that forms a channel from the cytoplasm into the nucleus (Fig. 1.2). Each nuclear pore is composed of about 30 different types of proteins called *nucleoporins*. Approximately 16 copies of all these proteins are assembled to form 16 columns that correspond to 8 large structures that are visible in routine electron micrographs.[2] The nucleoporins regulate traffic into or out of the nucleus. Messenger RNA molecules traversing the pores are attached to proteins that are recognized by pore proteins and which are guided outwards. All proteins found within the nucleus are originally synthesized within the cytoplasm; these proteins contain a string of basic amino acids that constitute a Nuclear Localization Signal, and which are also recognized by nucleoporins and guided into the nucleus.

Nuclear pores are interconnected by a network of proteins that are most abundant along the inside of the inner nuclear membrane. These proteins, termed *lamins*, belong to a class of filamentous proteins called *intermediate filaments* (they are intermediate in diameter between thinner *actin* filaments and thicker *myosin* filaments). *Intermediate filaments*, unlike other elements of the cytoskeleton, are *non-labile*, that is, once they are formed, they are not readily disassembled. In addition to *lamin intermediate filaments*, all cells contain

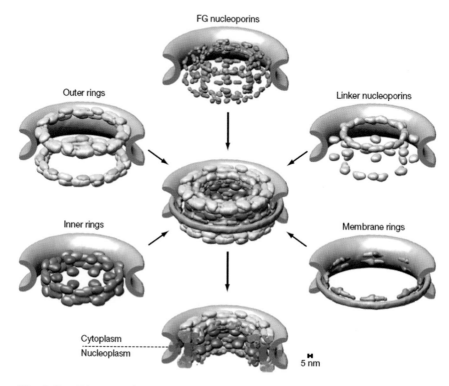

Fig. 1.2. Diagram of a nuclear pore, showing how the different types of nucleo-porins are assembled to form the entire structure. Reproduced from Ref. 2, with permission.

cytoplasmic *intermediate filaments*. These cytoplasmic filaments form tough, strong fibers that interconnect with each other at cell junctions and thus form a network of cables strung throughout the cell that give cells structural strength. Epithelial cells have intermediate filaments formed of *keratin*, connective tissue cells possess intermediate filaments formed from a related protein, *vimentin*, and muscle cells have filaments formed from *desmin* (see Ch. 2). Neurons have intermediate filaments called neurofilaments, and astrocytes possess intermediate filaments formed from *glial fibrillary acidic protein* (see Ch. 6).

There are four types of *lamin* proteins: A, B1, B2, and C. The nuclear meshwork of *lamins* appears to lend strength and stability to the nuclear envelope and plays a role in the overall shaping of the nucleus. This can be shown by experiments involving developing sperm cells. Sperm cell nuclei are very much smaller and more arrow-like in shape than most nuclei, which is in accord with their function of swimming as streamlined rockets through fluids of the reproductive tract. They also possess a type of *B lamin* that is not found in any other cell. If an ordinary cell is forced to produce sperm cell *lamins*, its nucleus will become deformed to resemble the pointed morphology of a sperm cell nucleus.[9]

The types of *lamins* produced by other cells seem to correlate with how specialized the cell is. For example, the unspecialized (stem) cells of the early embryo produce *B-type lamins* but not A- or C-types. Once cells become more specialized, they synthesize *A-type lamins.*[6] Also, a mutation in *A-type lamins* leads to a rare but devastating syndrome called progeria, in which cell nuclei become more fragile and teenaged patients acquire the aged appearance and fragile skins and bones of elderly people.[25] It is likely that variations in nuclear shape between different cell types involve rearrangements in nuclear *lamin* proteins.

Another subset of proteins closely associated with *lamins* are proteins of the so-called nuclear matrix. The nuclear matrix forms a scaffold upon which the chromatin of chromosomes is organized. The nuclear matrix contains proteins like *topoisomerase*, which can cause bending of the DNA double helix, and anchoring proteins like *SATB* (*special AT-rich sequence binding protein*, which binds to DNA). The DNA of nuclear chromosomes is attached to the nuclear matrix at an estimated 30,000 to 80,000 spots (called matrix attachment regions or *MARs*). Each of these attachment spots is not always active; when activated, a MAR will cause a loop of chromatin to be pulled closer to the nuclear matrix. This appears to be one mechanism for the activation of cell-specific genes: mature lymphocytes possess more MARs than immature lymphocytes, and if the MARs are inactivated, the genes normally active in mature lymphocytes will not be transcribed.[22]

The chromatin attached to the nuclear matrix contains the DNA and the DNA-associated proteins found in each chromosome. Human cells contain 23 pairs of chromosomes, for a total of 46. Some of the chromosomes are rather long and contain 1,000 to 2,000 genes; other shorter chromosomes contain only several hundred genes. Chromosomes are not randomly tangled together, but occupy discrete territories at specific spots along the nuclear envelope. Also, the gene-rich portions of chromosomes tend to be located at the center of the nucleus, whereas the gene-poor regions are found adjacent to the inner nuclear membrane.[7]

It may seem puzzling to speak of "gene-poor" and "gene-rich" regions of chromosomes. What are chromosomes if they are not linear arrays of genes? In fact, only a small amount (1.5%) of the DNA in chromosomes actually can be read, or transcribed, into messenger RNA molecules that code for proteins. The human genome contains only about 25,000 such gene sequences. The remainder of the DNA (almost one meter long, if all the chromosomes are laid end-to-end) contains highly repeated sequences that do not code for proteins at all.[3]

The DNA of each chromosome is basically composed of two intertwined strands of repeated sugar molecules (deoxyribose sugars), plus associated bases. In some ways, DNA is not really a remarkable molecule: cells commonly make a large variety of polymers of sugar molecules. For example, long chains of sugars called *glucose-amino-glycans* (*GAGs*) are commonly secreted by cartilage cells, mast cells, and epithelial cells. Other polymers of sugar (glycogen, starch) are used as a storage form of nutrients by cells and accumulate within the cell cytoplasm. What makes DNA unique is that the double helix of deoxyribose sugars contains within it sequences of bases (adenine, thymine, cytosine, and guanine, or A, T, C, and G) that form a code for the assembly of proteins.

Regulation of DNA transcription

How is the code on the DNA for a gene converted to a molecule of mRNA for transport out of the nucleus? Messenger RNA molecules are created by an enzyme, *RNA polymerase II*, that reads the code on a gene. This enzyme needs the assistance of many other proteins to accomplish this, however. Each gene sequence is preceded by a short

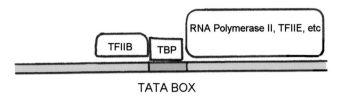

TATA BOX

Fig. 1.3. Simplified diagram showing how accessory proteins link RNA polymerase II to the promoter region of a gene. Adapted from Ref. 37.

DNA sequence, 25 bases "upstream," called the TATA box, which is located in the so-called promoter region of the gene and which is enriched in T and A nucleotides. A binding protein called the *TATA box binding protein* (*TBP*) first localizes to this TATA box and identifies where additional proteins should bind subsequently (Fig. 1.3). Accessory proteins for *RNA polymerase II*, called *Transcription Factors for Polymerase II* (*TFIIs*), perform functions like forcing open the DNA double helix so that the interior code can be read by *RNA polymerase*. With the aid of these proteins, RNA polymerase can read genes and produce a molecule of mRNA for export from the nucleus.

Even these proteins are not sufficient to ensure that a gene will be read. Additional DNA-binding proteins, which differ between one cell and another, are also needed. These DNA-binding proteins are called transcription regulating factors. They bind to DNA sequences called regulatory or enhancer sequences that are located hundreds or thousands of nucleotides distant from a gene on a particular chromosome. When transcription regulating proteins bind to enhancer regions, they cause the formation of a loop in the chromosome that brings the enhancer region close to the promoter region (Fig. 1.4). A complex of proteins binds to the enhancer region and helps stabilize the transcription factors located at the promoter. This action can enhance transcription by 1000-fold. Each cell in the body possesses its own mix of transcription regulating factors; this mix determines which genes will be transcribed for a muscle cell, a lymphocyte, a liver cell, etc.[3]

About 2000 different transcription regulating factors, amounting to almost 8% of the human genome, have been identified. Many of

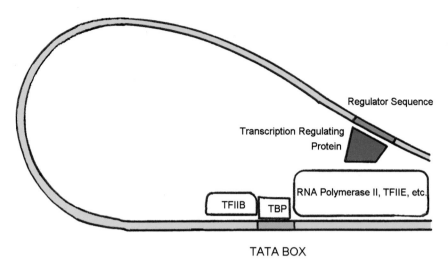

TATA BOX

Fig. 1.4. A simplified diagram showing how transcription regulating proteins, bound to a regulatory sequence on a chromosome, interacts with TFII proteins and RNA polymerase II at the promoter sequence of a gene (adapted from Ref. 37).

these proteins fall into a few basic categories. For example, some transcription regulating proteins belong to the so-called *"leucine zipper"* family.[1] Two of these proteins can bind to each other because they each have a stretch of peptides containing 4–5 leucine amino acids located at equal distances from each other. When two *"leucine zipper"* proteins approach each other, the complementary stretches of leucines interlock, much like the teeth of a zipper (Fig. 1.5).

Subsequently, the two interlocking proteins settle down on the DNA, aided by the positively charged amino acids at the DNA binding regions of the proteins that are attracted to the negatively charged phosphate molecules that form links between deoxyribose sugars of the DNA. Since the two proteins form mirror images of each other, they bind to a DNA sequence called a "palindrome." In a palindromic sequence, each half of the sequence is also a mirror image of the other half, e.g., a sequence of nucleotides such as AATTGC-CGTTAA is a palindrome. Many transcription regulating proteins, such as the *C/EBP* family (see the passages in Chapter 4 that describe the generation of fat cells), belong to the leucine zipper family.

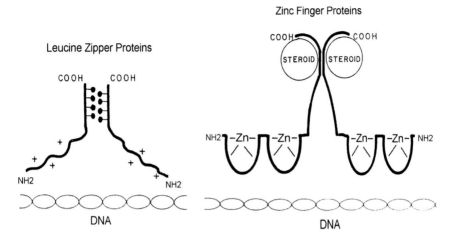

Fig. 1.5. Simplified diagrams of two types of transcription regulating proteins. Both leucine zipper proteins and zinc finger proteins form dimers that bind to palindromic DNA sequences. Adapted from Ref. 1.

Other transcription regulating proteins belong to the *"zinc finger"* family of proteins (Fig. 1.5). These proteins have regions of amino acids that bind zinc and are thrown into long loops, or *zinc fingers*. The shape of the loop, plus the positive charge associated with the zinc atom, allow the loops to interact with palindromic sequences of DNA. Important examples of this family of proteins are the receptors for steroid and thyroid hormones, which affect cell function by modulating transcription.[16,18,37]

Other families of transcription regulating proteins exist. Some proteins, for example, have two DNA-binding helices that are joined by a straight portion of the protein. These are termed *helix-turn-helix* proteins, and include the very important *homeotic* proteins that will be discussed in detail in Chapter 2. A final family is the so-called *STAT* family (for Signal Transduction and Activator of Transcription). This protein family will be examined in Chapter 8. The molecular structures of most of these transcription regulating proteins is becoming known in some detail.[18]

Considering the length of nuclear DNA and all the proteins associated with it, it almost seems surprising that a cell's genetic material would have room to fit inside a tiny cell nucleus. In fact, however, we must remember that even if a strand of chromosomal DNA is very long, it is also very delicate and thin. It has been calculated that if all of a cell's DNA were tightly compacted into a ball, it would occupy only 17% of the volume of the nucleus.[7] So there seems to be room enough, after all.

One thing about chromosome anatomy that *is* very remarkable is that a chromosome shrinks in length tremendously when it is compacted in preparation for cell division and mitosis. For example, chromosome number 22 shrinks in length from 1.5 centimeters to about 2 micrometers, a 10,000-fold reduction in length![3] How is this accomplished?

Most investigators feel that chromosomes are compacted for mitosis by a series of hierarchical coiling operations (Fig. 1.6). The DNA double helix itself forms the first level of coiling; the second level of coiling occurs when DNA becomes wrapped around roughly disc-shaped core particles composed of *histone* proteins. These core particles, in turn, form another coiled arrangement rather like the cord on a telephone. The larger chromatin coils are attached at intervals to proteins of the nuclear matrix. When the nuclear envelope breaks down during mitosis, it might be that the nuclear matrix proteins themselves can form an even larger coil. One could construct a model of a chromosome by stretching out part of a large "slinky" toy until it was straight and then fastening a telephone cord to the "slinky." In the interphase condition, the slinky would remain straight and coils of chromatin would hang off from one side; in the mitotic condition, the slinky would snap back to its original coiled condition and smaller coils of chromatin would radiate away from the slinky "axis." There is some evidence that this appealingly simple model has some similarity to reality, but it may not be strictly true in detail.[17,29]

One other variable that is noteworthy between the nuclei of different types of cells is that some cells have a prominent nucleolus, while others do not. The nucleolus represents the junction of genes for ribosomal RNA that are located on different chromosomes. When active, its function is to assemble new ribosomal subunits out of ribosomal RNA plus associated proteins. Cells that synthesize many

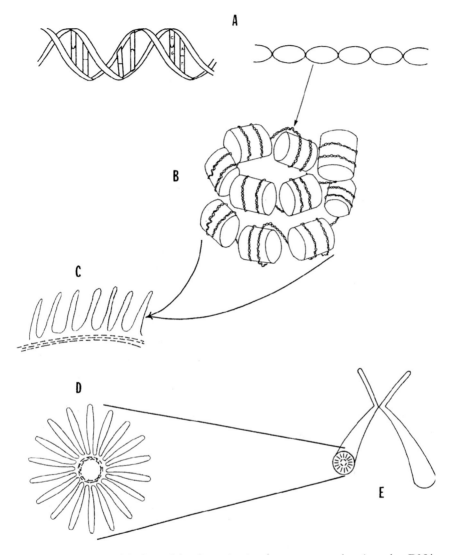

Fig. 1.6. A simplified model of a mitotic chromosome, showing the DNA double helix (a) wrapped around core particles made of histone proteins to form a coil (b), which is thrown into longer loops attached to the nuclear matrix (c). During mitosis, the nuclear envelope breaks down, allowing the nuclear matrix proteins to detach from the nuclear envelope and to form an axial core of the chromosome (d). Loops of chromatin are detectable in fully condensed, mitotic chromosomes.

proteins require lots of ribosomes, and thus tend to have enlarged, active nucleoli (see, for example, the large nucleoli present in oocytes and in nerve cells (Figs. 2.4 and 6.1)).

From the data reviewed above, it is clear that a cell biologist can infer a lot about a cell simply by looking at the morphology of its nucleus: a cell with dispersed chromatin and a large nucleolus will likely be transcribing many thousands of genes, whereas a cell with condensed, dark-staining chromatin probably transcribes only the minimal number of genes needed to keep the cell alive, plus a few genes for its particular specialty. The basic specializations of cells are the topics of the next chapter.

2. Regulation of the Endoplasmic Reticulum

When new functions of a cell arise, driven by changes in nuclear gene transcription, organelles of the cytoplasm must be adjusted to meet the new demands of the nucleus. The positions, activity, and volumes of each organelle may change, depending upon the specialized function of the cell. How do these changes come about? The differentiation of a plasma cell from a lymphocyte illustrates particularly well the challenges involved in answering this question.

Plasma cells are cells that secrete *immunoglobulins*, proteins (antibodies) that selectively bind to foreign molecules (antigens) that have invaded the body. When an antigen is presented in precisely the right way to a lymphocyte by a so-called antigen presenting cell, the lymphocyte is stimulated to differentiate into a plasma cell. The lymphocyte begins life as a small cell with a marginal amount of cytoplasm; as it turns into a plasma cell, the volume of the cytoplasm expands considerably and becomes filled with massive amounts of rough endoplasmic reticulum (rER) and Golgi stacks (Fig. 1.7).

This is appropriate for the function of the plasma cell. Plasma cells secrete prodigious amounts of protein (180,000 molecules per hour!) at about six times the rate of other secretory cells like fibroblasts.[14] In order to secrete so much protein, plasma cells must make appropriate adjustments to their protein-making machinery.

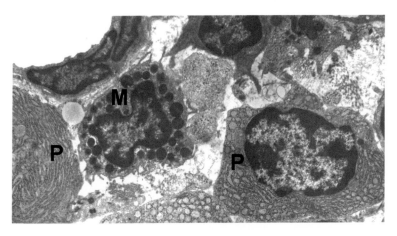

Fig. 1.7. A transmission electron micrograph showing two plasma cells (P) and a mast cell (M). The cytoplasm of the plasma cells is filled with masses of parallel stacks of rER. Courtesy of Dr. Lee Leak, Dept. Anatomy, Howard University College of Medicine.

The steps in synthesizing a protein are familiar elements of an introductory biology class:

(i) a molecule of messenger RNA (mRNA) is transcribed from the DNA in the nucleus,
(ii) the mRNA is modified by editing and splicing and is then exported from the nucleus, where it binds to cytoplasmic ribosomes,
(iii) ribosomes "read" the sequence of bases on the mRNA and, by binding the corresponding tRNA-amino acid complex, translate this sequence into a chain of amino acids, and
(iv) if the resulting protein is destined for export from the cell, it has a "signal sequence" of amino acids on one end that is threaded through the membrane of the rER so that it winds up within the lumen (cisterna) of the rER.

From this point, the completed protein will be packaged into membrane-bound vesicles that bud off from the rER and move to the Golgi stacks so that the protein can be further modified (addition of carbohydrate molecules, for example). Finally, vesicles will then leave the Golgi stacks to fuse with the plasma membrane and deliver their protein contents into the environment outside of the cell.

How is traffic between these organelles coordinated for secretion of proteins? The positions of the rER and Golgi stacks, plus the traffic between them, are regulated by the array of microtubules that sprout from the centrosome, near the nucleus (Fig. 1.8). Microtubules, hollow structures that resemble miniature drinking straws, grow from the centrosome, so that their "plus" ends wind up near the cell membrane and their "minus" ends remain at the centrosome. Microtubules function as a "railroad track" system that delivers membrane-bound structures to specific destinations within a cell. The membranous organelles travel along these "railroad tracks" with the aid of so-called motor proteins that act as tiny locomotive engines and move their cargoes down the tracks.

Golgi membranes acquire *dynein* motor proteins that drag the Golgi stacks down towards the centrosome by traveling towards the

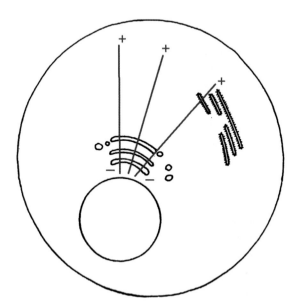

Fig. 1.8.   Microtubules (blue lines) radiate from the centrosome and organize Golgi stacks in a position near the nucleus.

"minus" ends of microtubules. This can be demonstrated experimentally: if Golgi membranes are isolated from disrupted cells and then injected into a healthy cell, they will immediately attach to microtubules and migrate to the centrosome. So, the reason why the Golgi apparatus is located close to the nucleus is becoming clear. Likewise, vesicles that bud off of the rER and which contain secretory proteins also acquire *dynein* motors that automatically deliver their contents to the Golgi stacks for further modifications. Membranes of the rER appear to travel in the opposite direction upon microtubules and thus wind up in parts of the cell farther from the Golgi apparatus.[33] If microtubules are experimentally disrupted in cells, the Golgi membranes become dispersed all over the cell, whereas the rER membranes collapse towards the cell nucleus.

If microtubules play a major role in determining the positioning and trafficking between the Golgi and rER, what anchors the microtubules to a position near the nucleus? This fundamental question has been answered in cells from roundworms (*Caenorhabditis elegans*). A protein called *Zyg12* attaches the cloud of proteins surrounding the centrioles (centrosome) to the nuclear envelope.[15] So it appears that a hierarchy of binding proteins and motor proteins determine the positions of many cellular organelles.

When plasma cells secrete massive amounts of immunoglobulins, it becomes necessary to increase the size of their rER by almost four-fold compared to the rER of lymphocytes. In a certain sense, this is unsurprising: if the genes for immunoglobulins are activated and large numbers of mRNA molecules are generated for translation into protein, it is appropriate to up-regulate rER and Golgi membranes. But how do these organelles "know" that there is a greater need for them and what could cause such a remarkable growth of the rER to produce the masses seen in plasma cells? The answer to this question is somewhat complicated and involves an event called the unfolded protein response.[13,19,36,38] This response takes place when the rate of the synthesis of a protein exceeds the rate at which the protein can be processed. There are several steps in the response.

First, when an mRNA molecule binds to a ribosome that settles down onto the rough ER, a growing polypeptide chain (in this case,

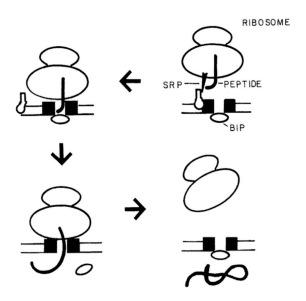

Fig. 1.9.   Diagram of a translocon pore, which threads a new protein into the rER and which is blocked at one end by the *BIP* protein. Initially, the signal sequence in a growing polypeptide chain is bound to the signal recognition particle (SRP) and then drawn into the translocon pore. After complete synthesis of the polypeptide chain, the signal sequence is cleaved off, the protein folds into the proper shape, *BIP* once again attaches to the translocon pore, and the ribosome detaches from the cytoplasmic surface of the rough endoplasmic reticulum (modified from Ref. 13).

an immunoglobulin) emerges from the ribosome and is fed into the cisterna of the rER via a membrane pore called the translocon (Fig. 1.9). The translocon is composed of a protein called *Sec61* that forms a channel through the membrane. Once the newly synthesized immunoglobulin is injected into the rER, the translocon is re-sealed on the inside by another protein called *BIP* (this strange name comes from the fact that the protein was discovered due to its ability to *Bind Immunoglobulin Proteins*). *BIP* does not only serve to seal up the translocon pore, but also has the ability to bind to hydrophobic portions of proteins that have not folded properly within the rER. Other so-called chaperone proteins will then facilitate the proper folding of rER proteins. This normally works just fine.

The unfolded protein response

If too much mRNA message is directed towards the rER and too much protein accumulates within the rER, the proteins overwhelm the ability of chaperone proteins to help fold them. Hence, misfolded proteins accumulate within the rER and saturate binding sites on the BIP protein. When this happens, the BIP protein dissociates from a membrane protein called *inositol requiring enzyme 1 (IRE1)*.

This minor event actually has a galvanizing effect on cell function. When the *IRE1* proteins form dimers of 2 proteins each, the proteins then can autophosphorylate each other and activate their function. The *IRE1* protein dimers then act as a peculiar type of enzyme that

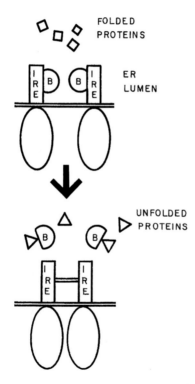

Fig. 1.10. When *BIP* (B in the figure) binds an excess of misfolded proteins, it is released from the rER membrane, permitting the dimerization of *IRE1* proteins (modified from Ref. 19).

specifically splices out a 26 nucleotide long intron loop out of the mRNA for another protein called the *X-box binding protein* (*XBP*). Introns are extraneous portions of the mRNA sequence that are usually removed in the nucleus prior to the export of the mRNA. Removal of the intron allows the mRNA for *XBP* to finally be translated into an active protein. But this *XBP* protein is only a single protein. How could it be responsible for such dramatic effects upon a cell's structure?

The basis for the power of the *XBP* is that it is a DNA-binding transcription regulating factor. When it binds to DNA, it binds to a type of DNA sequence called the X-box and stimulates the transcription of at least two dozen genes for proteins that are required for assembling the rER (these include *Sec61, signal sequence receptor, signal peptide protease, signal recognition particle*, a number of chaperone proteins, etc.). This leads to the synthesis of all the components of the rER and an expansion of the rER that is needed for the secretion of so much protein.

This sequence of events explains how, when the rER fills up with too much protein, it sends a signal to the nucleus to generate more rER.

The *XBP* protein is not the only one required for the transformation of a lymphocyte into a plasma cell. Another protein with the unusual name of *BLIMP* (*B-lymphocyte induced maturation protein*) is also needed for B-lymphocytes to turn into plasma cells. *BLIMP* is another transcription regulating protein (a zinc finger protein) that suppresses the activation of about 250 genes that are active in lymphocytes and which code for things such as B cell membrane receptors. So, *BLIMP* is turned on to suppress the features of B-lymphocytes, and then *XBP* is turned on to activate the features of plasma cells.[26]

This whole process is not only useful in explaining the transformation of lymphocytes into plasma cells, but very likely applies to other cell transformations as well. For example, the pancreas begins its existence as a modest sprouting of ducts and tubules from the lining of the primitive gut. Duct cells are simple, undistinguished cells with a very simple morphology. But when they mature into pancreatic acinar cells and begin secreting pancreatic enzymes, they too acquire massive amounts of rER, probably by the same mechanism as described above. So, an examination of the process governing plasma cell formation provides a useful insight into the specialized features of other cells.

This scheme shows how the volume of the rER might be adjusted, but it provides no information about how the structure of the rER is formed. Both smooth and rough endoplasmic reticulum membranes form either tubules or flattened, parallel sheets; neither configuration is natural for lipid bilayers, which spontaneously form only spherical vesicles when dispersed in water. Something must be forcing the ER membranes into the configurations seen in cells, but what is it? A number of intrinsic ER proteins (*reticulon* or *DP1* proteins) appear to be good candidates for supplying such a force: when these proteins are mixed with lipids, they form hollow tubules.[11,27,31] Similarly, proteins called *Golgins* or *CLASP* proteins may be responsible for the architecture of the Golgi apparatus.[5]

3. Regulation of Mitochondrial Number

Many cells of the body (e.g., cartilage cells, fibroblasts) live quiet lives within masses of extracellular material. Such cells have low requirements for energy, do not intensely utilize the energy-rich compound, *ATP*, and possess few mitochondria. Other cells, such as kidney cells that spend huge amounts of energy to recover molecules from the urine or muscle cells that steadily contract, require lots of *ATP* and possess many mitochondria to produce it. How are numbers of mitochondria adjusted to meet the energy requirements of a cell?

This question can be studied in a number of ways. One approach, for example, is to expose a cell to *thyroxine*. Two to three days after an injection of *thyroxine*, tissues from a rat will display about a two-fold increase in oxidative metabolism, primarily due to increased numbers of mitochondria. In confirmation of this, the mRNAs for mitochondrial proteins such as *Cytochrome C* and the *adenine nucleotide transporter* that imports the precursor for *ATP* into mitochondria are all increased by *thyroxine* administration. How do these events come to be?

Thyroxine acts by entering the cell nucleus and binding to *thyroxine receptor proteins*. Activated *thyroxine receptors* then bind to specific DNA sequences in the enhancer regions of genes called *thyroxine response elements*. Once in place, the *thyroxine receptor* also binds to another protein called the *thyroxine receptor associated protein*, which has the ability to activate *RNA polymerase* and influence transcription of the coding region of the gene for a protein. At least 54 genes are known to be influenced by *thyroxine*. So it would seem logical that *thyroxine* could simply turn on the genes required for mitochondrial biogenesis. However, here we come to a roadblock on our way to a solution: genes coding for mitochondrial components such as *Cytochrome C* or *adenosine transporter* **lack** *thyroxine* response elements![35,37] So how could they be turned on by *thyroxine*?

The answer to this problem appears to be that *thyroxine* activates the gene that codes for a protein called *nuclear respiratory factor-1* (*NRF-1*). This is a DNA-binding protein that regulates the activity of over 400 genes, many of which control overall cell metabolism.[16] Among the genes activated by *NRF-1* are the genes required for mitochondrial components. So, *thyroxine* may partly act upon mitochondria *indirectly*, by stimulating the appearance of a transcription regulating protein that controls their number.

Once mitochondrial proteins are synthesized in the cytoplasm, how do they reach mitochondria? Each mitochondrial protein has a signal sequence at its N-terminal end that is recognized by translocator complexes of proteins that are present in the outer and inner mitochondrial membranes. These feed the required proteins into the interior of the mitochondrion.[3,16]

Thyroxine is not the only stimulus that causes the appearance of new mitochondria within cells. Mitochondrial genesis can also be regulated by a specialized protein that serves as a "fuel sensor" within cells. This protein is called *AMP-activated protein kinase* (*AMPK*). Simply put, this is an enzyme that becomes active when energy in a cell becomes depleted and the *AMP/ATP* ratio increases (e.g., as seen during starvation or fasting). In response, this enzyme can suppress the synthesis of fatty acids, proteins, and glycogen, and increase glucose uptake and glycolysis to replenish the energy supplies of a cell. In addition, *AMPK* also activates *NRF-1*, and so can stimulate the genesis of new mitochondria. Thus, there are several mechanisms that transform the energy economy of a cell and regulate mitochondrial number. This explains a number of remarkable events in cells. For example, repeated exercise can transform mitochondria-poor white muscle fibers into mitochondria-rich red muscle fibers (the type of muscle cells seen in "dark meat" of postural muscles). This is due to a stimulation of *AMPK* by exercise and energy depletion of the muscles.[10,16,24]

Once new mitochondria are formed, they can be directed to the parts of a cell that have the greatest need for energy. If living cells are viewed under a microscope, it is easy to see that mitochondria are in continual movement. This is because they travel along the network of microtubules in cells, just as other organelles do. The motor proteins that move mitochondria in this way all require *ATP*. If they drag a mitochondrion into an energy-poor region of the cell, the motor proteins "run out of gas" and strand the mitochondrion in this area. This is actually a good thing, because the immobile mitochondrion will then release enough *ATP* in the region to both replenish local energy stores and also to get moving along the microtubules once more. The proteins responsible for this ingenious scheme have recently been identified: the motor protein *kinesin* binds to an adaptor protein called *milton*, which in turn regulates the movement of mitochondria by binding to a protein of the outer mitochondrial membrane called *Miro*.[32]

Mitochondria do not only differ in number between types of cells, they also differ in morphology. The inner mitochondrial membrane of all mitochondria is folded into structures called cristae. However, all cristae are not created alike. In most cells, cristae appear as flattened, shelf-like folds within mitochondria. In cells that make steroid hormones, the cristae take on a tubular shape. In some astrocytes in the brain, cristae form perfect triangles or prisms in cross-section (Fig. 1.11)![8] No one knows why these morphological changes take place, or how they are related to the specialized functions of mitochondria in specialized cells. Recently, some mitochondrion-specific proteins have been shown to be involved in these transformations. Enzymes found on the outer mitochondrial membrane called *mitofusin* or *OPA1* regulate the fusion and fission of mitochondria. Proteins within mitochondria such as *mitofilin* appear to regulate the structure of cristae.[12] More work needs to be done to explain how these proteins are involved in changes in mitochondrial structure.

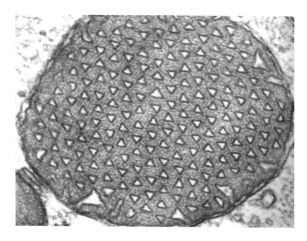

Fig. 1.11.   Electron micrograph of an astrocyte mitochondrion, showing prismatic cristae. Reprinted from Ref. 8, with permission.

The mechanisms described above do not always operate smoothly. One extreme example of a dysfunction of these systems is found in oxyphil cells of the parathyroid gland. These cells, apparently derived from the hormone-secreting chief cells of the gland, become almost entirely filled with hundreds of mitochondria that leave little room for other organelles. No explanation has yet been found for why these cells fail to utilize the mechanisms described above to regulate their numbers of mitochondria.

4. Control of Centrioles and Cilia

Centrioles are very peculiar cell organelles. They are composed of dimers of *tubulin*, like most cytoplasmic microtubules, but instead of forming single, long, hollow tubules, the *tubulin* proteins are combined to form long triplets of tubules that are fused together. Nine of these triplets form the core of a centriole (Fig. 1.12). Each barrel-shaped centriole is positioned rigidly at a 90-degree angle relative to its nearby "daughter" centriole and the pair of centrioles is in turn surrounded by a cloud of proteins called the pericentriolar matrix that makes up the remainder of the centrosome.

Fig. 1.12.   Each barrel-shaped centriole is composed of 9 triplets of microtubules.

In cells that are dividing, each pair of centrioles duplicates once during every cell cycle, so that all daughter cells possess only a single pair of centrioles. Some post-mitotic epithelial cells, however, have the capacity to generate *hundreds* of centrioles in sub-membrane structures called deuterosomes. These centrioles then move towards the plasma membrane to form basal bodies, structures from which cilia grow (Fig. 1.13).[30] Clearly, then, one function of centrioles is to promote cilia formation. Several proteins, such as *SAS-6* and a transcription factor called *Foxj1*, appear to be essential for this function. Cilia are much more complicated that the simple bundles of microtubules that they appear to be. Recent estimates suggest that about 300–500 proteins may be required for full ciliary function.[23]

What do centrioles do in non-ciliated cells? They are a component of the centrosome, and the centrosome is vital for the generation of microtubules that form the mitotic spindle, so it could be that centrioles are needed for the coordination of chromosomal movement. However, a number of cell types, such as mature oocytes, have diminished centrosomes and *absent* centrioles, and nevertheless move chromosomes during meiosis just fine.[21] Thus, centrioles might be dispensable for this function.

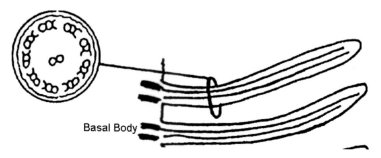

Fig. 1.13.   Diagram showing how basal bodies of centrioles form the origin of the axoneme of microtubules within a cilium.

One investigator, Guenter Albrecht-Buehler, has proposed a novel function for centrioles. He has noted that the dimensions of centrioles are perfect for the refraction of certain wavelengths of light (infrared frequencies). Also, the geometry of the centrioles, located at 90 degrees to one another, could, at least in theory, provide some information as to the direction of a light source. Albrecht-Buehler has shown that cells in culture do orient towards pulses of infrared light, and if they are "blinded" by shining a light on the centrosome, they lose this ability.[4] So, possibly, centrioles may function as some sort of rudimentary "eyes" for cells.

This proposal is not as outlandish as it may seem at first. Cilia have now been shown to have sensory functions in many cells, including kidney cells, neurons, and olfactory cells. In the kidney, single, non-motile cilia appear to monitor the volume of urine flowing through a tubule. If this sensory function is lost by a mutation in specific ciliary proteins, tubule cells react as if urine flow is blocked, and greatly enlarge to form the cysts found in polycystic kidney disease. Non-motile sensory cilia in the brain affect neuronal function; if their function is disturbed, a neurological disorder called Bardet-Biedl syndrome occurs.[28] Most notably, the outer segments of photoreceptor cells of the retina are basically highly modified cilia. Conceivably, a sensory function for centrioles and cilia that is rudimentary in most cells has been adapted and expanded to promote the function of specific sensory cells.

It is possible that centrioles functioned in ancient, ancestral cells as some sort of an intracellular sensor. Subsequent movement of duplicated centrioles to the cell surface could have resulted in a single, sensory cilium, and then, further duplication of cilia and the acquisition of motility in cilia resulted in the type of cilia seen in pseudostratified ciliated columnar epithelia that use cilia to move mucus.

5. Control of Overall Cell Shape

All of the material discussed above shows how we are gaining an understanding of how cells control the amounts and shapes of their internal organelles. However, how is the overall shape of the cell itself governed?

The cell membrane alone is too fragile to control cell shape: composed of an equal mix of membrane proteins and phospholipids, the plasma membrane is efficient in forming a watertight film around a cell, but is very fluid and weak as a soap bubble. To corral within the cell all of its components and to regulate cell shape, the cell membrane must be reinforced by attachments to cytoskeletal fibers of *actin*. This is accomplished by a network of filamentous proteins found on the cytoplasmic surface of the plasma membrane. One of these proteins, *spectrin*, forms connections between an anion transporter protein called *Band 3* that provides for the passage of bicarbonate ions through the plasma membrane.

Spectrin (illustrated in Fig. 1.14) also forms attachments to *actin* filaments via proteins called *ankyrin* and *tropomodulin*. *Tropomodulin*, in particular, seems essential for the interactions between the cytoskeleton and the plasma membrane. If it is deleted from tall columnar epithelial cells *in vitro*, the layer of *spectrin* on lateral cell membranes fails to form properly, and the cells are converted from tall cells to short cells.[34] Similar types of events must be occurring in most cells to ensure that cells retain an overall shape appropriate for their function.

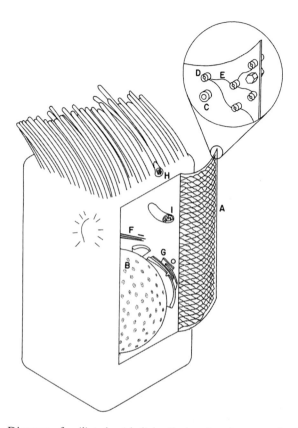

Fig. 1.14. Diagram of a ciliated epithelial cell, showing the network of proteins on the cytoplasmic surface of the plasma membrane (A). D labels the anion transport channels interconnected with each other by strands of spectrin (E). Other transporter proteins (C) permit the passage of water soluble molecules (sugars, amino acids, ions) across the plasma membrane, while still other integral membrane proteins (hexagons) function as receptors for extracellular ligands. The nucleus, with numerous nuclear pores, is seen at B. G and I label the Golgi apparatus and a mitochondrion, respectively. F indicates a bundle of actin filaments that is participating in the deformation of the plasma membrane. H represents a basal body of a cilium.

References

1. Abel T, Maniatis T (1989) Action of leucine zippers. *Nature* 341: 24–26.
2. Alber F, *et al.* (2007) The molecular architecture of the nuclear pore complex. *Nature* 450: 695–699.

3. Alberts B, *et al.* (2008) *Molecular Biology of the Cell.* NY: Garland Science.

4. Albrecht-Buehler G (1994) Cellular infrared detector appears to be contained in the centrosome. *Cell Motility Cytoskeleton* 27: 262–271.

5. Barr FA, Short B (2003) Golgins in the structure and dynamics of the Golgi apparatus. *Current Opinion in Cell Biol* 15: 405–413.

6. Constantinescu D, *et al.* (2006) Lamin A/C expression is a marker of mouse and human embryonic stem cell differentiation. *Stem Cells* 24: 177–185.

7. Dehghani H, Dellaire G, Bazett-Jones DP (2005) Organization of chromatin in the interphase mammalian cell. *Micron* 36: 95–108.

8. Fawcett DW (1981) The Cell. WB Saunders Co.: Philadelphia, p. 457.

9. Furukawa K, Hotta Y (1993) cDNA cloning of a germ cell specific lamin B3 from mouse spermatocytes and analysis of its function by ectopic expression in somatic cells. *Embo J* 12: 97–106.

10. Hardie DG, *et al.* (2003) Management of cellular energy by the AMP-activated protein kinase system. *FEBS Letters* 546: 113–120.

11. Hu J, *et al.* (2008) Membrane proteins of the endoplasmic reticulum induce high-curvature tubules. *Science* 319: 1247–1250.

12. John GB (2005) The mitochondrial inner membrane protein mitofilin controls cristae morphology. *Molec Biol Cell* 16: 1543–1554.

13. Johnson AE, van Waes MA (1999) The translocon: a dynamic gateway at the ER membrane. *Annu Rev Cell Dev Biol* 15: 799–842.

14. Kao WW, Berg RA, Prockop DJ (1977) Kinetics for the secretion of procollagen by freshly isolated tendon cells. *J Biol Chem* 252: 8391–8397.

15. Kemp CA, *et al.* (2007) Suppressors of *zyg-1* define regulators of centrosome duplication and nuclear association in *Caenorhabditis elegans*. *Genetics* 176: 95–113.

16. Klinge CM (2008) Estrogenic control of mitochondrial function and biogenesis. *J Cellular Biochem* 105: 1342–1351.

17. Kireeva N, *et al.* (2004) Visualization of early chromosome condensation: a hierarchical folding, axial glue model of chromosome structure. *J Cell Biol* 166: 775–785.

18. Liljas A, *et al.* (2009) Textbook of Structural Biology. World Scientific Publ.: Singapore, p. 248.

19. Liu CY, Shroder M, Kaufman RJ (2000) Ligand-independent dimerization activates the stress response kinases IRE1 and PERK in the lumen of the endoplasmic reticulum. *J Biol Chem* 275: 24881–4885.
20. Martin (2005) Archaebacteria (Archaea) and the origin of the eukaryotic nucleus. *Curr Opinion Microbiol* 8: 630–637.
21. Manandhar G, Schatten H, Sutovsky P (2005) Centrosome reduction during gametogenesis and its significance. *Biol Rep* 72: 2–13.
22. Ottaviani D, *et al.* (2008) Anchoring the genome. *Genome Biol* 9: 201.
23. Pazour GJ, Agrin N, Leszyk J, Wiotman GB (2005) Proteomic analysis of a eukaryotic cilium. *J Cell Biol* 170: 103–113.
24. Rockl KSJ, *et al.* (2002) Skeletal muscle adaptation to exercise training. AMP-activated protein kinase mediates muscle fiber type shift. *Diabetes* 56: 2062–2069.
25. Rodriguez S, *et al.* (2009) Increased expression of the Hutchinson-Gilford progeria syndrome truncated lamin A transcript during cell aging. *Eur J Hum Genet* 17: 928–937.
26. Shaffer AL, *et al.* (2004) XBP1, downstream of Blimp-1, expands the secretory apparatus and other organelles, and increases protein synthesis in plasma cell differentiation. *Immunity* 21: 81–93.
27. Shibata Y, Voeltz GK, Rapaport TA (2006) Rough sheets and smooth tubules. *Cell* 126: 435–439.
28. Singla V, Reiter JF (2006) The primary cilium as the cell's antenna: signaling at a sensory organelle. *Science* 313: 629–634.
29. Strukov YG, Belmont AS (2009) Mitotic chromosome structure: reproducibility of folding and symmetry between sister chromatids. *Biophys J* 96: 1617–1628.
30. Vladar EK, Stearns T (2007) Molecular characterization of centriole assembly in ciliated epithelial cells. *J Cell Biol* 178: 31–42.
31. Voeltz GK, *et al.* (2006) A class of membrane proteins shaping the tubular endoplasmic reticulum. *Cell* 124: 573–586.
32. Wang X, Schwarz TL (2009) The mechanism of Ca2+-dependent regulation of kinesin-mediated mitochondrial motility. *Cell* 136: 163–174.
33. Watson P, *et al.* (2005) Coupling of ER exit to microtubules through direct interaction of COPII with dynactin. *Nature Cell Biol* 7: 48–55.

34. Weber KL, Fischer RS, Fowler VM (2007) Tmod3 regulates polarized epithelial cell morphology. *J Cell Science* 120: 3625–3632.

35. Weitzel JM, Alexander K, Iwen H, Seitz HJ (2003) Regulation of mitochondrial biogenesis by thyroid hormone. *Exp Physiol* 88: 121–128.

36. Wiest DL, *et al.* (1990) Membrane biogenesis during B cell differentiation: most endoplasmic reticulum proteins are expressed coordinately. *J Cell Biol* 110: 1501–1511.

37. Yen PM, Ando S, Feng X, Liu Y, Maruvada P, Xia X (2006) Thyroid hormone action at the cellular, genomic and target gene levels. *Molec Cell Endocrinology* 246: 121–127.

38. Zhang K, Wong HN, Song B, Miller CN, Scheuner D, Kaufman RJ (2005) The unfolded protein response sensor IRE1α is required at 2 distinct steps in B cell lymphopoiesis. *J Clin Invest* 115: 268–275.

Chapter 2

HOW DO CELLS OF THE FOUR BASIC TISSUES ARISE FROM EMBRYONIC STEM CELLS?

The preceding chapter presented some of the more recent answers to the question of how a cell controls the amounts of its organelles to accommodate new demands for protein synthesis and energy. It is clear that we are beginning to understand how a cell, once it specializes to perform a function, can modify its anatomy appropriately. However, an even more basic question is how a cell becomes specialized in the first place.

About 200 different types of cells can be identified in the body; each cell type has its own specific job, such as secreting specific proteins into its environment, contracting to move other structures, etc. This bewildering variety of cell types can be somewhat simplified, however. All of the body's cells fall into only four basic categories, or tissues. These are:

1) epithelial cells, which form layers to cover exposed surfaces or line hollow organs,
2) muscle cells, which contract to move other structures,
3) cells of the nervous system, which provide a communications network for the body, and
4) cells of connective tissue, which "fill the spaces" in the body not occupied by the other cell types and which produce extracellular molecules that fight disease or which provide structural strength to organs.

Why do cells become specialized to form these four basic tissue types?[24] This has long been the million-dollar question in biology.

In order to answer this question, we must take a brief excursion into the field of embryology and development. It is not necessary, or even feasible, to attempt to even briefly summarize all the recent progress in the cell biology of embryology in this chapter. However, a small subset of developmental data can be employed to better understand how cells become specialized. The answers to this basic question are becoming increasingly clear for lower organisms such as flies or frogs. Somewhat surprisingly, it is only mammalian development that is still shrouded in mystery and controversy. How do the four basic tissue types develop in most animals, and do mammals share some of the mechanisms found in other organisms?

In many animals, a rudimentary "blueprint" for assembly of an embryo exists as far back in development as the stage of an unfertilized egg cell (oocyte). Examples of these blueprints are found in oocytes of flies (*Drosophila melanogaster*) or in oocytes of frogs (*Xenopus laevis*).

1. Cell Specialization in Drosophila

Unfertilized fly oocytes are ovoid structures with an apparently unremarkable anatomy (Fig. 2.1). However, hidden within them is a gradient of the mRNA for a DNA-binding protein called *Bicoid*, which is termed an egg-polarity protein. *Bicoid* protein accumulates at the anterior pole of the egg. *Bicoid* belongs to a family of DNA-binding proteins (*homeotic* proteins or *hox* proteins). Homeotic proteins all have a similar stretch of 60 amino acids that can bind to DNA; the portion of the gene that codes for this DNA-binding region of the protein is called the *homeobox*. Each *hox* protein binds to a specific DNA sequence called a hox response element and thus can regulate transcription of DNA.

At the posterior pole of the oocyte, a gradient of another protein, *Nanos*, forms. *Nanos* is an RNA binding protein that can regulate the translation of mRNA. Thus, these 2 maternally derived proteins define the anterior and posterior poles of the egg. Also, local activation of a membrane protein called *Toll* defines the ventral surface of the egg. It is clear that even before fertilization the egg is divided up into specific regions.

restricted to the mesoderm.[26] They will eventually play major roles in determining the identity of body segments, just as in the fly.

Finally, the mesoderm itself has a variable influence upon the remaining cells of the animal pole. Some of these cells will specialize into an ectoderm layer that forms the skin of the frog. The remaining animal pole cells will specialize into ectodermal cells that form the central nervous system (CNS) of the frog. The early CNS is formed when the ectodermal cells fold towards each other and form a hollow tube called the neural tube. This only happens because of an antagonism between two signaling molecules. One signaling molecule is called *Bone Morphogenetic Protein* (BMP) and is synthesized throughout the mesoderm. Other signaling molecules called *Chordin* or *Noggin* are synthesized in only a special region of the embryo called the organizer. Antagonism of the action of *BMP* by *Chordin* or *Noggin* prevents cells from acquiring a mesodermal fate and promotes the development of the neural tube and eventually, the central nervous system. So, the basic signals that cause the appearance of the four basic types of tissue — epithelium, muscle, connective tissue, and nerve — are beginning to be well understood in the frog.

3. Cell Specialization in Mammals

The main goal of this book is to explore how cells acquire their specialized forms and functions in the human (mammalian) body. So, why have we spent so much time covering the development of flies and frogs? The main reason is to put into context current controversies in mammalian embryology. Until very recently, mammalian development has been considered to be fundamentally *different* from the development of all other animals.

Unfertilized mammalian eggs appear to be basically unpolarized, unlike eggs of other species. After fertilization, mammalian eggs divide into equal-sized cells that form a blackberry-like structure called a morula. Eventually, the morula acquires a fluid-filled, central cavity and at this stage, the embryo is called a blastocyst (Fig. 2.3). A group of rounded cells called the inner cell mass forms at one end of the blastocyst, while at the opposite end, cells lining the central

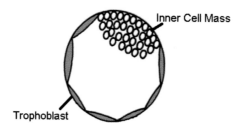

Fig. 2.3.   Diagram of a mammalian blastocyst, composed of cells of the inner cell mass and cells of the trophoblast.

cavity become flattened and form a layer called the trophoblast. This layer will eventually form the placenta, whereas the inner cell mass forms the embryo itself. In mammals, unlike in any other organisms, the first choice an embryonic cell must make is to become part of the placenta or part of the embryo. Only mammals create placentas.

Experiments on early mammalian embryos led to the conclusion that, unlike in most animals, mammalian embryogenesis is highly regulated. By this it is meant that early blastomeres are completely unspecialized and can potentially form any part of the embryo. And indeed, complete embryos can be generated from only one blastomere of the 2- or 4-cell embryo, and removal of a single blastomere does not seem to disrupt development. Thus, since the undivided egg appeared unpolarized, and since blastomeres of the early embryo did not seem to exhibit specialized properties, the idea of a pre-existing "blueprint" in the unfertilized egg that dictates the construction of an embryo seemed not to apply to mammals, unlike flies or frogs.[17]

Other investigators have challenged this conventional viewpoint. They argue that embryos are flexible and can indeed recover from the experimental removal of a blastomere, but that flexibility does not rule out the possibility of an orderly arrangement of blastomeres before such an experiment. New studies have shown that there are hints of asymmetry in unfertilized rodent eggs: clusters of mitochondria, proteins such as *STAT-3*, and clouds of dark-staining material have been found to localize to one pole of an oocyte[2,16,27] (Fig. 2.4).

A number of *hox* genes are expressed in mammalian oocytes.[10] One such *hox* protein, called *Cdx2*, appears to be localized at one pole

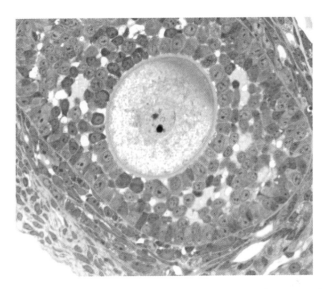

Fig. 2.4. Dark-staining material can occasionally be detected at one pole of unfertilized rat oocytes.

of unfertilized oocytes, and only 1 of 2 early blastomeres incorporates this protein into its cytoplasm. *Cdx2* then stimulates cells to acquire a trophoblast cell fate.[4] Finally, a detailed study of the cell fates of early blastomeres suggests that blastomeres are not randomly distributed throughout the embryo, but contribute to trophoblast or inner cell mass in a way dictated by their original positions relative to the animal or vegetal poles of the oocyte.[2]

It is clear that this argument is not yet resolved, but the possibility remains that the developmental theme of oocyte polarity, seen in frogs and flies, may yet apply to the development of mammals, as well. This is just another example showing that biology is not just a dry list of facts to be memorized, but an assortment of conjectures that hopefully approximates reality and which is continuously modified.

After the early stages of development, the inner cell mass, for completely unknown reasons, begins to change. Initially, all the cells of the inner cell mass appear to be able to generate any type of cell of the embryo. This highly adaptable ability is termed pluripotency, and the cells that exhibit this remarkable flexibility are termed pluripotent stem cells. Numerous studies have shown that if these cells are removed from the early embryo and cultured in the presence of appropriate signaling molecules, they can become muscle cells, bone cells, nerve cells, etc. This plasticity, however, is soon lost.

By about the second week of development in humans, the inner cell mass begins to divide into layers. The top layer forms the amniotic membrane. Second and third layers form sheets of epithelial cells called the epiblast and hypoblast, respectively (Fig. 2.5). Transformations of the epiblast are responsible for generating all the types of tissues that are eventually seen in the adult.

The critical changes in the epiblast layer take place at a visible groove that appears in the midline of the disc of the epiblast. The reasons for the formation of this groove, called the primitive streak, are still not known. Epiblast cells on either side of this primitive streak march towards the midline and then dive beneath the epiblast to form new cell layers. Initially, cells insert themselves into the hypoblast and shove the hypoblast cells aside to form a new layer called the endoderm, which, like in frogs, will form the gut of the embryo.

In the next step, more cells dive into the primitive streak from the epiblast and completely lose their epithelial character. Such cells migrate into the space between the epiblast and endoderm and form the connective tissue-like layer called the mesoderm. To form these migrating, freely wandering cells, the epiblast cells cease making the *cadherin* proteins that cause them to adhere to one another. They also

Fig. 2.5. The inner cell mass of a mammalian blastocyst forms a bilaminar disc containing the epiblast and hypoblast.

cease making the *laminin* proteins that form the basal lamina. In the absence of a basal lamina barrier beneath the epithelium, cells are freed to migrate. The exact pathways that cause cells to lose epithelial characteristics and acquire those of the mesoderm are not known with certainty; one cause is the activation for a gene for a *hox* protein called *Txl-2* after exposure to *BMP*.[22]

At specific locations along the primitive streak, mesodermal cells make decisions to form muscle cells, connective tissue cells, blood cells, etc. The proteins that switch on these decisions are gradually becoming known. For example, by exposing epiblast to the following sequence of signaling molecules, blood-forming mesodermal cells are produced[15] (Fig. 2.6).

Something similar to this sequence, verified *in vitro*, must be happening *in vivo* in the embryo. How these signaling molecules trigger the activation of some genes and the suppression of most of the

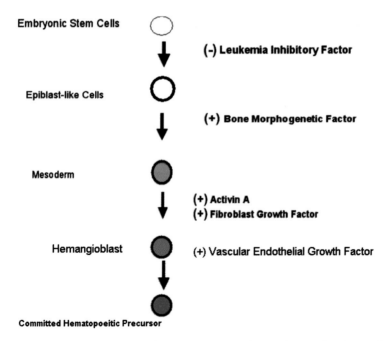

Fig. 2.6. Sequential exposure of stem cells to specific signaling molecules produces a specific cell fate.

remaining genes has not been elucidated with certainty. However, it is known that the signaling molecules that stimulate the appearance of blood-forming cells also activate the transcription of *hox* proteins specific for these cells.[21] So, thus far, some of the reasons for the formation of gut epithelium, connective tissue, blood cells, and muscle cells have been defined. However, what about the epidermis and nervous system, which form from ectoderm?

Epiblast cells that move towards the primitive streak but fail to dive into it form the final germ layer, the ectoderm. Ectodermal cells can remain on the surface of the embryo and form the epidermis. Alternatively, they can form yet another groove on the dorsal surface of the growing embryo and create a fold that eventually deepens into a tube called the neural tube (Fig. 2.7).

The epithelia-like cells lining the neural tube undergo rapid mitotic divisions and slowly migrate laterally away from it. As they do so, they lose some of the features of epithelia and acquire some of the properties of neurons and glial cells.

Another critical feature of the neural tube is that it must change from a mass of epithelial cells that lacks blood vessels — a characteristic of all epithelia — to a tubular structure of neuroepithelial cells

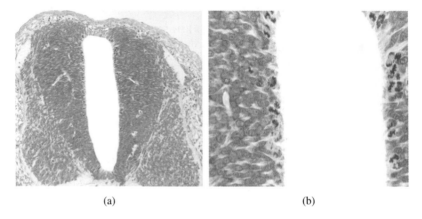

(a)   (b)

Fig. 2.7.   a. Low magnification view of the neural tube, formed from infoldings of the ectoderm. Clumps of sensory neurons that will form the dorsal root ganglia can be seen on each side of the neural tube. b. Higher magnification view, showing numerous, dark mitotic figures of neuroepithelial cells lining the lumen.

that are nourished by blood vessels. This is accomplished when meso-dermal angioblasts form clusters around the neural tube and then invade it with capillary sprouts. The reason for this development is the production by the neural tube cells of a protein called *Wnt* (named after a mutation in Drosophila called *Wingless*). The *Wnt* protein binds to receptors on mesodermal cells; these receptors activate a cyto-plasmic protein called *catenin*, which moves to the cell nucleus and regulates gene transcription. Mice with an impaired production of *Wnt* develop disorganized neural tubes lacking blood vessels.[19]

Once the neural tube fully forms, its cells must develop the spe-cialized features of nerve cells and abandon some of the features of epithelial cells. What are the specialized features of these two cell types?

4. Specialized Features of Epithelial Cells

A basic feature of mature epithelial cells is that they are polarized, i.e., they possess a basal plasma membrane that abuts the basal lamina, and an apical cell membrane that faces the lumen of a hollow structure. This polarity is essential for their function, so that they secrete basal lamina components in the right direction and apical secretion vesicles towards the lumen. Also, epithelial cells need to construct cell junc-tions, microvilli, and cilia at the apical rather than the basal region of the cell. But what causes this essential cell polarity?

Recent studies have shown that a family of proteins (called *PAR* proteins) is required for the generation of cell polarity. These proteins were first discovered in nematodes that had severe embryological defects, but since have been characterized in flies, frogs, and mam-mals. In fly eggs, they organize microtubules to point in one direction, and thus cause proteins like *Bicoid* or *Oskar* to be dragged along microtubules in different directions until they come to rest at the correct poles of the egg. In epithelia, *PAR* proteins are among the first to localize to the apical regions of a cell. Then, *PAR* proteins recruit proteins like *claudin* and *occludin* that form tight junctions, and also regulate other proteins that form intermediate junctions and desmosomes[8] (Fig. 2.8).

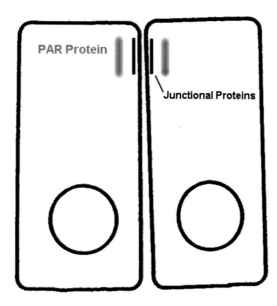

Fig. 2.8.   Diagram showing how *PAR* proteins recruit junctional proteins to the apical portion of an epithelial cell.

   Tight junctions (*zonulae occludens*), in particular, are important for this process, because they prevent diffusion of proteins in the apical cell membrane to other membranous surfaces of the cell. Intramembrane proteins of these junctions like *claudin, junction adhesion molecule-A* (*JAM-A*), and *occludin* cause cell membranes of adjacent cells to fuse with each other, forming water-proof seals that prevent the passage of fluid across a barrier of cells. Other proteins (*ZO* proteins) connect the membrane proteins with *actin* filaments of the cell skeleton. Tight junctions encircle the entire apical region of the cell and make a seal rather like the seal produced in sandwich bags. They also hinder movement of proteins within the cell membrane itself. So, after tight junction formation, proteins required for cilia, microvilli, and secretion granules can be addressed to apical membranes and will not diffuse from the apical region of the cell to incorrect locations in other regions of the cell membrane. This essential polarity allows for the construction of a typical epithelial cell[12] (Fig. 2.9).

wrap *myelin* around neuronal dendrites or around glial cell processes. Also, some axons become myelinated, and some do not. Recent studies suggest that the production of a signaling protein called *neuregulin* by a neuron determines how much *myelin* a glial cell will apply to an axon.[23]

Dendrites also have a critical influence on the function of neurons. The actual pattern of a dendritic tree in itself influences the distribution of electrical charge upon a neuron and can determine the firing pattern of a neuron.[25] Thus, the overall appearance of the dendritic tree of a neuron is not of mere academic interest, but is intrinsic to how the neurons of a given part of the brain function. In each brain region, nerve cells have shapes of dendritic trees that tend to be characteristic for that region (Fig. 2.11).

The factors that determine the eventual morphology of the dendritic tree are uncertain; a number of *homeotic* proteins called *Cut* and *Knot* have been shown in Drosophila to control the shape of dendrites by regulating the microtubules and *actin* filaments of the cytoskeleton.[11] This would be consistent with the ability of homeotic proteins to guide the development of specific brain and spinal cord segments during embryogenesis and to specify the characteristics of nerve cells within each segment.[1]

Knob-like structures called dendritic spines also influence neuronal activity by determining how many synapses affect the function

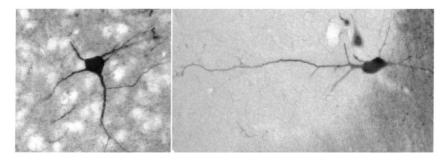

Fig. 2.11. Neurons stained for *calbindin* to illustrate their dendritic trees show a branching pattern that is highly variable between different cells. Thinner, more delicate axons are not visible in this figure.

of the cell. Another *PAR* protein, *PAR6*, is required for dendritic maturation and the acquisition of dendritic spines.[8]

Finally, yet another feature of nerve cells that they share with epithelial cells is that nerve cells secrete molecules (neurotransmitters) into their environment (synapses), but typically do not release these molecules unless they are specifically stimulated to do so. This type of secretion is called regulated secretion, as opposed to constitutive (continual) secretion. Cells that undergo regulated secretion store their secretory molecules in vesicles prior to secretion, and then in response to a stimulus move the storage vesicles towards the plasma membrane for fusion with it and the exocytosis of vesicle contents. The steps required for neurotransmitter secretion are beginning to be worked out.

The first step is that neurotransmitter is synthesized near the neuronal nucleus and packaged in vesicles that attach to micro-tubules via a motor protein called *kinesin*. This motor protein utilizes *ATP* to move the vesicles towards the plus ends of the microtubules that terminate far away from the cell body near synapses at the ends of an axon. Once in place, the vesicles are poised to fuse to the plasma membrane upon command. What, then, is the command (Fig. 2.12)?

When a wave of depolarization (action potential) flashes down the length of an axon, it affects the function of voltage-sensitive calcium channels located in the presynaptic membrane of the synapse. These channels open, allowing calcium to rush into the neuron. This acti-vates a vesicle-associated protein called *synaptotagmin*, which appears to function as a calcium sensor. *Synaptotagmin* is then able to activate other vesicle proteins called *v-SNAREs*. The activated *v-SNAREs*, in turn, can then interact with their targets: *SNARE* proteins on the presynaptic membrane (*t-SNARES* or *target SNARES*). All of these proteins combine to form a complex that drags the vesicle down onto the presynaptic membrane and cause the fusion of the vesicle mem-brane with the plasma membrane.[3]

This cellular machinery for neurotransmitter release can be selectively interfered with by a toxin produced by a bacterium called *Clostridium botulinum*. This botulism toxin, or botox, is now

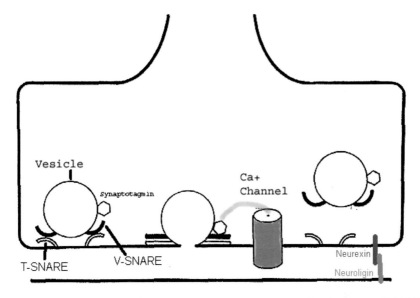

Fig. 2.12. Diagram of a synapse, showing synaptic vesicles bearing *v-SNARE* proteins, a voltage-dependent calcium channel that activates *synaptotagmin*, and adhesive proteins (*neuroligin* and *neurexin*).

commonly used to prevent the release of neurotransmitters at myoneural junctions, resulting in a smoothing of the skin.

All the details of this process of neurotransmitter release are still not completely clear. Also, although many of these proteins were discovered in nerve cells, they subsequently have been found in exocrine and endocrine epithelial cells that also undergo a regulated type of secretion.

One curious feature of regulated secretion is that the shape and size of secretory vesicles is somehow arranged to correspond to the specific content of a vesicle. For example, in neurons, excitatory neurotransmitters like *acetylcholine* or *norepinephrine* tend to be located in small, spherical vesicles, whereas the vesicles that contain peptide neurotransmitters are larger. Similarly, in pituitary endocrine cells, each hormone tends to be packaged within a secretion granule of a specific size: growth hormone is stored within rather large granules, whereas luteinizing hormone is stored within smaller granules, even though these protein hormones have rather similar molecular weights and sizes. The reasons for this variability in vesicle size are unknown.

In summary, although the shape of neurons differs from that of epithelial cells, both cell types have much in common: they are both polarized, they both attach to adjacent cells by utilizing similar proteins, and they secrete in a regulated manner. These shared features reflect the common developmental origin of nerve and epithelial cells.

6. Special Features of Connective Tissue and Blood Cells

Cells of connective tissue in many ways are simpler than epithelial cells or neurons, and appear to have lost some of the specialized features of these two cell types. Two main types of cells — fibroblasts and fat cells — are native to connective tissue, whereas the remainder of connective tissue cells — plasma cells, macrophages, lymphocytes, and mast cells — originate in the blood or bone marrow and are directed to migrate into connective tissue. Regardless, most of these cell types differ from epithelia and neurons in that they are not highly polarized. Also, these cells have abandoned the molecules that allow for cell–cell adhesion, and so are suspended, unattached to each other, within the abundant extracellular matrix that they secrete. Most of these cells, moreover, secrete their molecules from all surfaces of the cell more or less continuously and exhibit constitutive secretion.[24]

Fibroblasts, plasma cells, and macrophages secrete (among other things) *collagens* and *proteoglycans, immunoglobulins,* and *cytokines* like *tumor necrosis factor,* respectively. However, large secretory vesicles are not visible in these cells, since they do not store these molecules for any length of time, and instead secrete them immediately after they are synthesized. The one exception to this rule is the mast cell, which produces very prominent secretory granules filled with *heparin, proteoglycans,* and *proteinases* that are only secreted in response to the binding of an antigen onto molecules of *IgE* retained upon the cell membrane.

White blood cells share most of the features of other connective tissue cells, except that these cells *do* exhibit a regulated type of secretion and do store their molecules in secretory vesicles.

7. Special Features of Muscle Cells

Muscle cells are additional descendents of mesodermal cells, and in some ways occupy a position midway between epithelial and connective tissue cells. Muscle cells are unpolarized and generally lack secretory vesicles and in these ways resemble connective tissue cells. On the other hand, cardiac muscle cells do retain the ability to adhere to one another and can make cell junctions similar to those seen in epithelia. The differentiation of muscle cells will be discussed in a Chapter 3.

8. Germ Cells

Mouse and human oocytes, like those of frogs, express a *DAZ* protein, but in mammals, this protein seems to be uniformly distributed within the oocyte. Primordial germ cells in mammals seem to arise in the epiblast in response to *BMP* signaling from the extraembryonic ectoderm; these cells continue to express *DAZ*, whereas in the remainder of the embryo, cells lose the ability to express this protein. So, there are both similarities and differences between frogs and mammals in the process of generating germ cells.[6]

References

1. Akin ZN, Nazarali AJ (2005) *Hox* genes and their candidate downstream targets in the developing central nervous system. *Cell Molec Neurobiol* 25: 697–705.
2. Bischoff M, Parfitt D-E, Zernicka-Goetz M (2008) Formation of the embryonic-abembryonic axis of the mouse blastocyst: relationships between orientation of early cleavage divisions and pattern of symmetric/asymmetric divisions. *Development* 135: 953–962.
3. Butz B, Okamoto M, Südhof TC (1998) A tripartite protein complex with the potential to couple synaptic vesicle exocytosis to cell adhesion in brain. *Cell* 94: 773–782.

4. Deb K, *et al.* (2006) Cdx2 gene expression and trophectoderm lineage specification in mouse embryos *Science* 311: 992–994.

5. Fields IC, *et al.* (2007) v-SNARE cellubrevin is required for basolateral sorting of AP-1B–dependent cargo in polarized epithelial cells. *J. Cell Biol.* 177: 477–488.

6. Fox MS, *et al.* (2007) Intermolecular interactions of homologs of germ plasm components in mammalian germ cells. *Dev Biol* 301: 417–431.

7. Garber K (2007) Autism's cause may reside in abnormalities at the synapse. *Science* 317: 190–191.

8. Goldstein B, Macara IG (2007) The PAR proteins: fundamental players in animal cell polarization. *Developmental Cell* 13: 609–622.

9. Houston DW, *et al.* (1998) A Xenopus DAZ-like gene encodes an RNA component of germ plasm and is a functional homologue of *Drosophila boule*. *Development* 125: 171–180.

10. Huntriss J, Hinkins M, Picton HM (2006) cDNA cloning and expression of the human NOBOX gene in oocytes and ovarian follicles. *Mol Hum Reprod.* 12: 283–289.

11. Jinushi-Nakao S (2007) Knot/Collier and Cut control different aspects of dendrite cytoskeleton and synergize to define final arbor shape. *Neuron* 56: 963–978.

12. Laukoetter MG, Nava P, Nusrat A (2008) Role of the intestinal barrier in inflammatory bowel disease. *World J Gastroenterol* 14: 401–407.

13. Matusek T, *et al.* (2008) Formin proteins of the DAAM subfamily play a role during axon growth. *J Neurosci* 28: 13310–13319.

14. Mowry KL, Cote CA (1999) RNA sorting in *Xenopus* oocytes and embryos. *Faseb J* 13: 435–445.

15. Pearson St, Sroczynska P, Lacaud G, Kouskoff V (2008) The stepwise specification of embryonic stem cells to hematopoietic fate is driven by sequential exposure to Bmp4, activin A, bFGF and VEGF. *Development* 135: 1525–1535.

16. Pepling ME, Wilhelm JE, O'Hara AL, Gephardt GW, Spradling AC (2007) Mouse oocytes within germ cell cysts and primordial follicles contain a Balbiani body. Proc Natl *Acad Sci USA* 104: 187–192.

17. Rivera-Perez JA (2007) Axial specification in mice: ten years of advances and controversies. *J Cell Physiol* 213: 654–660.

18. Slack J (2001) Essential developmental biology. Ch. 13. Oxford: Blackwell Science Ltd.
19. Stenman JM, *et al.* (2008) Canonical Wnt signaling regulates organ-specific assembly and differentiation of CNS vasculature. *Science* 322: 1247–1251.
20. Südhof TC (2008) Neuroligins and neurexins link synaptic function to cognitive disease. *Nature* 455: 903–909.
21. Taghon T, *et al.* (2002) *HOX-A10* regulates hematopoietic lineage commitment: evidence for a monocyte-specific transcription factor. *Blood* 99: 1197–1204.
22. Tang SH, *et al.* (1998) The Tlx-2 homeobox gene is a downstream target of BMP signaling and is required for mouse mesoderm development. *Development* 125: 1877–1887.
23. Taveggia C, *et al.* (2008) Type III Neuregulin-1 promotes oligodendrocyte myelination. *Glia* 56: 284–293.
24. Telser AG, Young JK, Baldwin KM. (2007) Elsevier's Integrated Histology. Mosby, Inc.: Philadelphia.
25. Vetter P, Roth A, Hausser M (2001) Propagation of action potentials in dendrites depends on dendritic morphology. *J Neurophysiol* 85: 926–937.
26. Wacker SA, McNulty CL, Durston AJ (2004) The initiation of *Hox* gene expression in *Xenopus laevis* is controlled by Brachyury and BMP-4. *Developmental Biol* 266: 123–137.
27. Young JK, Baker JH, Allworth A (1999) Evidence for polar cytoplasm/nuage in rat oocytes. *Anatomy and Embryology* 200: 43–48.
28. Zeng M, *et al.* (2008) DAZL binds to the transcripts of several *Tssk* genes in germ cells. *BMB Reports* 41: 300–304.

Chapter 3

HOW DO ADULT STEM CELLS CONTRIBUTE TO BASIC TISSUE FUNCTIONS?

The ability of cells to acquire specialized features — polarity, junctions with neighboring cells, secretion granules, altered amounts and distributions of major organelles, etc. — is crucial for the performance of specialized functions. However, specialization does not come without a price. By acquiring specialized features, many cells become committed to a future without cell division, which would often necessitate abandoning these hard-won specialized features to permit a division of a cell into two equal daughter cells. For example, an equal division of a mammoth fat cell into two equal daughter cells is just not feasible. Thus, a preservation of a mitotic ability for all cells would not make sense in terms of energy expenditure: it would be too costly to continually differentiate and then de-differentiate to perform mitosis. A post-mitotic future for most cells, however, is itself a problem.

Adult tissues are not static entities, but suffer damage or require alterations in functional capacity as they grow or adapt to changing metabolic conditions. To allow these adaptations, new cells are required. Where are they to come from if many adult cells are post-mitotic? The answer to this problem is to form a reservoir of relatively undifferentiated stem cells for tissues.

1. Embryonic Stem Cells

Stem cells can be defined as relatively undifferentiated cells that have a high capacity for cell division, and that yield daughter cells which either remain undifferentiated (forming a reserve of more stem cells) or which transform into more specialized, non-dividing cells (replacing other cells that have died). The most adaptable stem cells yet found are those derived from the inner cell mass of the embryo, which appear to have an almost unlimited ability to transform into any cell type in the body. Recent studies have shown that this remarkable plasticity can be induced in ordinary fibroblast cells by transfecting them with retro-viruses coding for only four factors, called *Oct3, Sox2, Klf4,* and *c-Myc.*[36] Somehow these proteins reprogram the genome of the fibroblasts and allow them to act as pluripotent stem cells. Completely pluripotent stem cells, however, are not thought to persist into adulthood.

2. Stem Cells of Epithelia

Stem cells can also be found in adult tissues. While they are not quite as adaptable as embryonic stem cells, recent work has shown them to be available and useful in all four of the basic types of tissues. For example, stem cells can be found in epithelia, particularly in stratified or pseudostratified epithelia formed from multiple layers of epithelial cells rather than from a single layer of cells (Fig. 3.1).

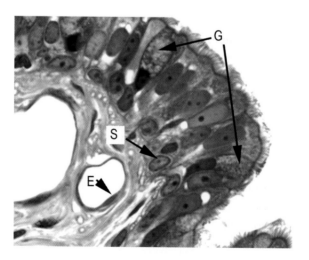

Fig. 3.1.   Rounded cells at the basal surface of a pseudostratified, ciliated, columnar epithelium act as stem cells (S). Other cells in the epithelium (goblet cells, G) secrete mucus, while the majority of cells are ciliated. Beneath this pseudostratified columnar epithelium, the simple squamous epithelial cells (E) of blood vessels can be seen.

In these epithelial tissues, the stem cells divide and then transform into the more specialized cells of the epithelium. In the respiratory system, daughter cells of the basal stem cells become transformed into mucus-producing goblet cells or ciliated cells that move the mucus up the respiratory tract towards the pharynx. Why do these cell transformations occur?

The ability of epithelial stem cells to divide appears to be dependent upon a transcription factor called *p63*. When mouse embryos are genetically modified to lack *p63*, they fail to form any stratified epithelia and have skins composed of only one layer of cells![15,32] They also lack the pseudostratified epithelia of the respiratory system. So, some of the signals responsible for generating epithelial stem cells are now known. But what causes the further differentiation of these stem cells into goblet cells and ciliated cells?

The limited data available on goblet cell differentiation suggests that a transcription factor called *Math1* first appears in developing goblet cells and stimulates the transcription of genes for mucin proteins.[35] The synthesized mucin proteins accumulate in characteristically large, pale-staining secretory granules that are directed towards the apical portion of the cell, resulting in the typical "goblet" shaped appearance of the cell.

Eleven human mucin proteins are known.[24] Some mucins, like salivary mucins, are readily soluble in water, whereas others form a thick gel upon epithelial surfaces. All mucin proteins are decorated by large numbers of carbohydrate molecules (mucin is 80% carbohydrate by weight). Since the carbohydrate molecules are highly charged, they repel each other, forcing the polypeptide chain of a mucin into a rigid, rod-like configuration. As a result, in solution, mucin proteins behave like a mass of logs floating in a river, getting tangled up with each other and causing mucus to be very viscous. Mucin proteins also have anti-bacterial properties; dying bacteria trapped in the mucus are carried from the respiratory system by ciliary action and destroyed in the gastrointestinal system.

In addition to mucins, the mucus secreted by goblet cells also contains a protein called *trefoil factor 3*.[18] This is one of a family of cloverleaf-shaped proteins that can bind to cell membrane receptors of adjacent cells. When it does so, it promotes the synthesis of a transcription factor called *Foxj1*. This factor, in turn, stimulates the migration of basal bodies to the apical cell surface and their elongation to produce 200–300 cilia per cell.[4,43] As these signals become better understood, cell biologists can go beyond a simple description of an epithelium towards a mechanistic description of how specific types of epithelia form. All of these changes are dependent upon the initial division and differentiation of epithelial stem cells.

Stem cells also play a role in repairing the damage to single layers of epithelial cells. The simple squamous epithelium (endothelium) of blood vessels, for example, undergoes damage in a process called atherosclerosis. The initial event in atherosclerosis is an accumulation of cholesterol bound to low-density lipoproteins just beneath the vascular endothelium. This appears to irritate the endothelial cells, which respond by secreting a protein into the bloodstream called *monocyte chemoattractant protein* (*MCP*).[3] Circulating blood cells called monocytes respond to this protein by migrating to the injured vessel, leaving the bloodstream, and transforming into phagocytic macrophages (called foam cells in atherosclerosis) that attempt to ingest the offending lipids. If this is not successful, the macrophages accumulate into a growing atherosclerotic plaque that deforms the interior of a vessel and eventually causes the endothelial cells to loosen and tear off from the interior of the vessel.

Loss of endothelial cells from a blood vessel during atherosclerosis can have serious consequences. When blood proteins are exposed to molecules of the connective tissue normally hidden from the blood by the endothelium, they will spontaneously initiate clot formation. In some circumstances, such as a wound to the skin, this will help minimize blood loss and aid in healing, but in atherosclerosis, clot formation can occlude blood vessels. In the heart, this can cause a myocardial infarction (heart attack); also, if clots break off from the interior of large blood vessels, they can migrate through the vasculature to cause a pulmonary embolism and death of portions of lung tissue or else can lodge in intracranial vessels and provoke a stroke. It is thus clear that repair of endothelium is of great importance. How is it accomplished?

Some degree of vessel repair can be carried out by movement or proliferation of endothelial cells immediately adjacent to the injury. However, the mitotic potential of adult endothelial cells is rather low and extensive damage cannot be repaired in this manner. Another means is via the recruitment of endothelial progenitor (stem) cells from the bone marrow. This is accomplished in a number of steps.

First, tissues and cells near a damaged vasculature experience a decrease in oxygen, which activates an oxygen-sensing protein called *hypoxia inducible factor 1* (*HIF1*). Second, cells activated by *HIF1* secrete a number of proteins.[9,10] One of them is called *stromal cell derived factor-1*, which is released into the blood and acts as a homing signal that attracts endothelial progenitors to leave the bloodstream at the correct place. Another protein is *vascular endothelial growth factor* (*VEGF*), which is carried by the bloodstream to the bone marrow and which can stimulate the production of endothelial progenitor cells by nine-fold within 24 hours.[6] The progenitor cells acquire endothelial markers like adhesive *cadherin* proteins, travel to the site of injury, and intercalate themselves into the injured vessel lining to help repair the damage. A better understanding of all the factors needed to activate these endothelial stem cells could be of great value in the treatment of cardiac ischemia, fractures, and burns.

3. Stem Cells of Muscle

A. *Skeletal muscle stem cells*

Stem cells, called satellite cells, also exist among the cells that constitute adult skeletal muscle. This is fortunate, since the elongated skeletal muscle fibers could not reasonably be expected to divide. Normally, nuclei of satellite cells amount to about 4–8% of the total population of cell nuclei within a muscle. They are difficult to distinguish from the other muscle cell nuclei; in electron micrographs, the nuclei of satellite cells appear to be more heterochromatic (darker staining) than those of muscle fibers and furthermore can be distinguished because they stain for specific proteins like *M-cadherin* and a transcription factor called *Pax7*.[2]

Satellite cells are normally relatively inactive, but upon injury to muscle, they are stimulated to multiply and replace the damaged muscle fibers. This reparative function of satellite cells is a daunting task. To perform it, these cells must replace skeletal muscle fibers that have an enormous size: each fiber has a diameter of about 0.1 mm and can extend to a length of 20 mm, making them several hundred-fold longer than they are wide. Each fiber may also contain as many as 500 cell nuclei, which never divide. How can a tiny cell like a satellite cell replace such a gigantic structure, and what initiates this transformation?

The first stage in muscle repair is the initial damage to muscle cells, called muscle fibers. Damaged fibers appear to release a number of molecules that attract macrophages and neutrophils to the area. These molecules include *monocyte chemoattractant protein (MCP)* and another protein called *keratinocyte-derived chemokine*.[34] Macrophages and neutrophils recruited to the area of damage begin phagocytizing the debris extruded from dying muscle fibers (Fig. 3.2).

In addition to removing damaged cell constituents, macrophages secrete growth factors like *leukemia-inhibitory factor* and *basic fibroblast growth factor* that stimulate the proliferation of satellite cells. Within 3 days after injury, damaged fibers will thus be surrounded by a mixture of inflammatory cells and multiplying satellite cells (Fig. 3.3).

Another important molecule is *hepatocyte growth factor (HGF)*, which was originally discovered to be present in the blood of rats undergoing regeneration of the liver. *HGF* is now known to also regulate muscle growth, and is released from damaged muscle fibers. This molecule is another stimulator of the proliferation of satellite cells.[2]

The next step, the differentiation of satellite cells into muscle-forming cells, or myoblasts, involves a number of DNA-binding proteins (transcription regulation factors). One of these is called *Pax7*. *Pax7* is one of nine proteins possessing a "paired box" of two DNA-binding regions. Resting satellite cells possess high levels of

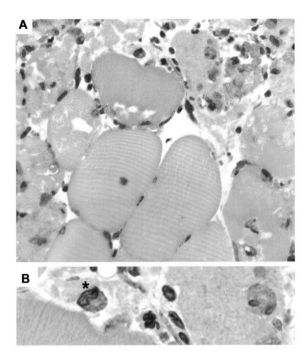

Fig. 3.2. A. Low magnification view of normal muscle fibers (bottom of Fig. 3.2A) and damaged muscle fibers surrounded by phagocytic cells (top of Fig. 3.2A). B. High magnification view of damaged muscle fibers, showing the bean-shaped nucleus of a macrophage (*).

Pax7, which seems to maintain the cells in an undifferentiated state. When the cells are stimulated to form myoblasts, *Pax7* transcription falls. This event appears to free up sites on DNA and allows for the binding of another transcription regulating protein called *MyoD*.[25] *MyoD* seems particularly important for muscle cell development and can stimulate the transcription of at least 36 muscle-specific genes.[5] These targets of *MyoD* include genes for *actin, myosin, titin,* and *troponin*. What is the function of these proteins in adult muscle cells?

The proteins of skeletal muscle have been the objects of curiosity for more than seventy years. Early histologists were able to discern a banding pattern composed of dark- and light-staining bands, called sarcomeres, crossing the narrow cytoplasm of a muscle fiber.

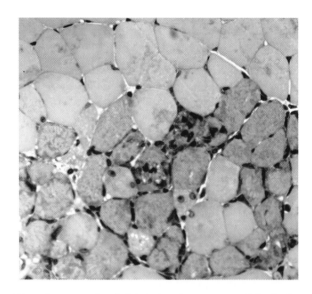

Fig. 3.3. After phagocytosis of debris has been accomplished, damaged muscle fibers (center of the figure) become infiltrated with proliferating satellite cells, which eventually will reform the fibers. Normal, undamaged muscle fibers, cut in cross section, surround the damaged fibers.

Dark-staining bands were called the A-bands because they exhibited a particular mode of transmitting polarized light (called Anisotropy). The lighter-staining bands were termed I-bands because they exhibited Isotropy under polarized light. Later, it was found that the I-bands contained a thin, dark-staining line. The German biologists who reported this termed this feature the *Zwischenlinie*, German for the "line between lines." Not long after this report, this feature was abbreviated as the Z line. Also, a lighter-staining region of the dark A-band was found and was named for the German word for "bright," or *hell*. Hence, the term for the H-band was born. At that point, however, the significance of these bands was unclear. It was only after more modern methods of microscopy and protein chemistry were devised that the arrangement of proteins within muscle fibers was understood (Fig. 3.4).

A diagram of a muscle fiber shows the orderly arrangement of proteins within the cell that enable it to contract. The I-bands are

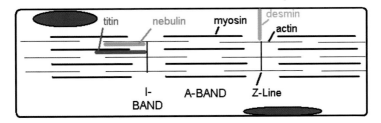

Fig. 3.4. Diagram of a skeletal muscle fiber, showing the proteins that are found within the A-bands and I-bands.

composed mostly of thin filaments of actin. These are attached at one end of the filament to a protein called *alpha-actinin* that composes the Z-line and which help anchor the *actin* filaments in place. The Z-line, in turn, is anchored to the cell membrane by the somewhat thicker filaments of a protein called *desmin* (desmin belongs to a class of cytoskeletal proteins called intermediate filaments). Another protein, called *nebulin,* is closely associated with the *actin* filaments and seems to serve as an intracellular "ruler," that prevents the addition of too many or too few *actin* monomers to the *actin* filament and which ensures that the thin filaments remain the right length.[20] Additional proteins like *troponin* and *tropomyosin* are associated with the *actin* filaments and regulate the ability of *actin* to interact with nearby filaments of *myosin.*

The A-bands of skeletal muscle are composed largely of thick filaments of myosin. *Myosin* filaments are closely associated with a very long protein called *titin.* This protein truly deserves its name, which is derived from the word for titanic. It has a molecular weight of almost four million, making it the largest single protein known in mammalian cells. It closely adheres to the filaments formed from *myosin* monomers and then extends all the way to the Z-line. *Titin* apparently functions to limit the amount of tension that develops within an A-band during muscle contraction (Fig. 3.5).[39]

What mechanisms account for this incredibly ordered, almost crystalline, array of proteins within a skeletal muscle cell? It partly can be attributed to the specific pattern of binding sites that these proteins utilize to adhere to each other. Also, another large protein that

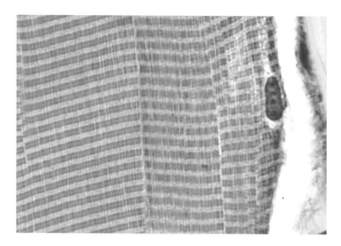

Fig. 3.5. High magnification view of a skeletal muscle fiber, showing the A-bands, I-bands, and the Z-line within the I-bands.

is related to *titin*, called *obscurin*, seems to have a role in initially organizing muscle proteins. *Obscurin* is expressed at the same time that *titin* begins to adhere to *myosin* during muscle differentiation, and if its expression is blocked experimentally, the A-bands, but not the I-bands, of muscle are disrupted.[1]

The function of some of these proteins has been relatively well understood for some time. Contraction of the muscle cell occurs when the *actin* and *myosin* filaments slide against each other. This sliding movement does not occur unless an electrical impulse travels into the cell and causes the release of calcium from the smooth endoplasmic reticulum. Calcium binds to a complex of proteins attached to *actin* filaments (*troponin*). In the presence of calcium, *troponin* releases its hold upon another *actin*-associated protein called *tropomyosin*. This change in the location of *tropomyosin* allows the *actin* filament to interact with and slide upon the surface of adjacent *myosin* filaments, generating the force for muscle contraction.

One final puzzle in the differentiation of satellite cells into adult muscle fibers is the question of how satellite cells line up and then fuse together to form a muscle fiber. The fusion of the plasma membranes of several cells to form a giant cell is otherwise a relatively rare event within the body. Membrane fusion appears to be mediated by a peculiar membrane protein called *ADAM* (*A Disintegrin and Metalloprotease*). This protein begins to be expressed in fusing satellite cells and appears to be responsible for the fusion of cell membranes during muscle regeneration.[17]

B. *Cardiac muscle stem cells*

Cardiac muscle cells differ somewhat from skeletal muscle cells in that they are much smaller and usually contain no more than two nuclei, rather than hundreds (Fig. 3.6).

For many years, it was widely accepted that another difference between skeletal and cardiac muscle cells was that stem cells for cardiac muscle were absent, so that a damaged heart had no real potential for repair. This viewpoint has been strongly challenged by a number of studies that have demonstrated the existence of cardiac progenitor cells in the normal myocardium and also in the bone marrow. Moreover, administration of *hepatocyte growth factor* and *insulin-like growth factor 1* promotes their proliferation and

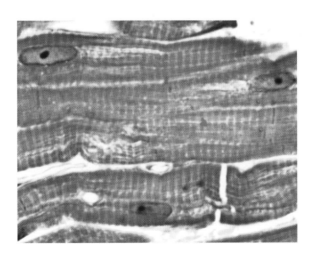

Fig. 3.6. High magnification view of cardiac muscle. Junctional regions at the ends of each muscle cell (one is visible here as a white line between one cell and another) are called intercalated discs.

differentiation, much like that of skeletal muscle satellite cells.[16,31] This newly discovered existence of cardiac muscle stem cells may provide new methods for the treatment of heart disease.[40]

In addition, administration of other growth factors can also stimulate the regrowth of blood vessels in damaged areas of the heart. *Fibroblast growth factor*, which stimulates the growth of endothelial cells, and platelet-derived growth factor, which stimulates the growth of vascular smooth muscle cells, have been shown to enhance vessel formation and cardiac function in pigs.[21] A renewed appreciation of the regenerative potentials of cells in the heart has been central to the development of new strategies in cardiac medicine.

Cardiac muscle cells develop under the influence of several heart-specific transcription factors, including homeotic proteins like *Nkx2–5* and *Isl1*. Cardiac muscle cell progenitors initially have the capacity to form the endothelial and smooth muscle components of the growing heart, as well as the myocardium.[7] This plasticity perhaps explains why cardiac muscle can occasionally be seen to replace the smooth muscle of vessels in the lungs (personal observation).

One curious feature of developing cardiac muscle cells is that they, unlike skeletal muscle cells, have the ability to undergo mitosis. This can even occur at times when sarcomeres are fully formed; surprisingly, the movement of mitotic chromosomes does not appear to fatally disrupt the highly ordered arrangement of *myosin* and *actin*.[8] This mitotic ability is lost in adult cardiomyocytes. If the presence of sarcomeres itself is not an obstacle to mitosis, what suppresses the ability of adult heart cells to undergo mitosis?

Mitosis is a landmark event in the life of most cells. It represents a period of dramatic rearrangement of nuclear DNA. Another such event, taking place prior to the separation of mitotic chromosomes, is the synthesis of new DNA. These two events mark two phases of the cell cycle, termed the M phase and the S phase, respectively. In between these two phases of nuclear transformation are two other phases called Gap phases (G_1 and G_2). The proteins that govern the transitions between each of these phases of the cell cycle are gradually becoming known.

For example, when rats are trained in a maze to locate food, the hippocampus is activated and its neurons stain positively for proteins required for learning.[28]

What is meant by learning, and how have learning-related proteins been identified? Learning occurs when a nerve cell generates new synapses that allow two previously unrelated stimuli to become associated with each other. For example, if a rat goes through a blue door in a maze to reach food, it will learn to associate the blue door with a reward. The neural circuits that recognize food and which detect the blue door thus acquire new connections that associate these two stimuli with each other. Some of the molecules required for this type of associative learning are becoming identified.

Two such memory-related proteins are called *CREB* (*Cyclic AMP-response Element Binding protein*) and an associated protein called *TORC* (*Transducer Of Regulated CREB*). These proteins detect when a neuron is simultaneously stimulated by two different stimuli. One stimulus, for example, could elevate levels of *cyclic AMP* in a neuron, while another stimulus could elevate intracellular levels of calcium. When both of these molecules are elevated, they are detected by *TORC*, which then phosphorylates its partner, *CREB*. Phosphorylated *CREB* migrates to the nucleus, where it settles down upon DNA and stimulates the production of another protein called *brain-derived neurotrophic factor* (*BDNF*). Finally, *BDNF* is released from the neuron and modulates the structure and number of nearby synapses.[42] These steps appear to be important for the modulation of neuronal function and the generation of new memories. After maze training, hippocampal neurons, but not neurons of other brain structures, stain positively for phosphorylated *CREB*.[28]

In rats, several thousand new hippocampal neurons are born every day, though only a fraction of these persist to become integrated into the hippocampus. Rates of neurogenesis in the adult hippocampus can be stimulated by moderate exercise or by training-induced learning. If a toxin for proliferating cells (a DNA methylating agent called methylazoxymethanol acetate) is given to rats, not only is the formation of new neurons specifically blocked, but associative learning is also diminished. Data from rats suggest that the birth of new neurons and the creation of new spatial memories are closely linked and dependent upon each other.

Indirect data also suggest that findings in rats apply to humans. Some subsets of humans are forced by their professions to learn vast amounts of data about their environment. One such subset is the London taxi driver, who is required to study for two years to learn the labyrinthine configuration of the city. Brain imaging studies have shown that these taxi drivers have an enlarged hippocampus, relative to the hippocampi of control subjects. Thus, learning appears to stimulate hippocampal neurogenesis in humans as well as rats.[22]

In addition to being a prime site of adult neurogenesis, the hippocampus has a number of other unique features. Cells of the CA region are particularly sensitive to stress or hypoxia, and will respond to corticosteroid hormones or hypoxia-induced oxidative stress with massive losses of neurons that are not seen in other less vulnerable brain regions.[26] Also, in Alzheimer's disease, the hippocampus is one of the earliest brain regions to show damage.[41] Damage to the hippocampus results in memory impairment and dementia that affect as many as one out of five Americans over the age of 75.

Many of the molecules that play a role in Alzheimer's disease have now been identified. A prime candidate for the damage seen in this disorder is a transmembrane protein called *amyloid precursor protein (APP)*. When it is cleaved by a number of enzymes (beta and gamma secretases and presenillins), it can yield a smaller peptide called the *Aβ peptide* that can form harmful aggregates called amyloid plaques. However, the identification of these molecules has not solved a number of basic mysteries about Alzheimer's disease and the hippocampus. For example, the mRNA for *APP* is distributed widely across the brain and is also present in regions that are never damaged in Alzheimer's disease, so why is the hippocampus specifically affected?[23] Also, what exactly is the function of *APP* in the normal brain?

One possible explanation for these questions is that the continual remodeling of neuronal structure in the hippocampus, due to localized learning-related neurogenesis, could somehow be related to the foci of damage seen in Alzheimer's disease. Perhaps, *APP* may normally contribute to the remodeling process seen in learning.[29]

A better understanding of the special features of the hippocampus will be required for a clinical solution to the problem of Alzheimer's disease. A basic defect in our understanding of the hippocampus is that no one knows why the environment of the hippocampus, which apparently is very different from the rest of the brain, favors neurogenesis in the first place.

5. Stem Cells of Connective Tissue

Connective tissue is composed of resident cells (fibroblasts or their specialized descendents, osteoblasts or chondrocytes, plus fat cells) and of non-resident cells that migrate into it from the bone marrow (white blood cells and lymphocytes). Subsequent chapters will discuss the proliferation and differentiation of bone marrow-derived cells. For now, a brief discussion of the proliferation of fibroblasts and fat cells would be appropriate.

Fibroblasts normally live quiet, non-reproducing lives nestled within the masses of collagen fibers that they secrete (Fig. 3.7). This non-mitotic state is enforced by the production of a number of proteins that were named on the basis of their molecular weights (*p15*, *p16*, *p18*, *p21*, *p27*, and *p57*). All of these proteins function as inhibitors of the *cyclin-dependent kinases* that drive cells through the cell cycle, so that fibroblasts normally do not divide.[27] However, when connective tissue and adjacent skin are damaged, as occurs after a wound, this inactive state changes. A wound in the skin promptly fills with blood from severed vessels, and a clot, composed of a meshwork of blood platelets and clotting proteins, temporarily seals the wound. In addition to forming a clot, platelets also deliver *platelet-derived growth factor* (*PDGF*) into the surrounding tissues. This and other growth factors release fibroblasts from mitotic inhibition and cause the multiplication and migration of fibroblasts to form scar tissue within the dermis.[19] Thus, unlike cells of other tissues, fibroblasts are fully competent to resume mitosis and do not require the aid of stem cells.

This competence of fibroblasts for mitosis has long been taken advantage of by biologists interested in the control of cell division.

In culture, fibroblasts readily divide until they reach the stage of about 50 population doublings; then, cell division ceases and the cultures become senescent. This surprising limit on cell division was first reported in the 1960's by Leonard Hayflick, and since has become known as the Hayflick limit. What causes this mysterious limitation in the number of cell divisions of fibroblasts?

Surprisingly, the Hayflick limit seems related to small decreases in the length of chromosomes that occur with each cell division. At the end of each chromosome, called a telomere, the normal double-stranded arrangement of DNA tapers into a single strand of 6 base sequences of DNA that are repeated thousands of times. In non-dividing cells, this single-stranded DNA is covered by binding proteins so that it is not destroyed by enzymes that correct DNA damage. When DNA is duplicated prior to mitosis, DNA polymerase enzymes cannot completely duplicate this single strand of telomeric DNA, so that for each cell division, 50–100 bases are lost from the telomeres. After enough cell divisions, the telomeres become so eroded that they attract the attention of another protein called *p53*. This protein constantly surveys nuclear DNA for errors or breaks; if *p53* detects such errors, it either stimulates minor DNA repair or else provokes the death of the cell. Thus, the gradual loss of telomeres seems to account for the mitotic senescence of fibroblasts in culture.

If fibroblasts are limited to a finite number of cell divisions, how can stem cells in general continue to replenish the cells of the body? *In vivo*, stem cells of many types appear to undergo as many as 1000 cell divisions without difficulty. How can stem cells escape doom due to telomere shortening? Stem cells, unlike most cells in the body, all seem to possess an enzyme called *telomerase* that regenerates the telomeres during each cell division. This, at least in part, explains the proliferative ability of stem cells. Also, an acquisition of telomerase is noteworthy in at least 85% of tumor cells, and helps explain why some types of cancer cells can become "immortalized" and divide indefinitely.[33]

The other cell type native to connective tissue, the adult fat cell, is post-mitotic and cannot divide. This fact led to considerable controversy some years ago: if fat cells don't divide, does this mean that the massive weight gain seen in severe obesity results solely from an enlargement of existing fat cells? This controversy has now subsided with the identification of stem cells for adipose tissue that multiply and provide for additional storage of fat when necessary. These cells resemble fibroblasts and adhere to the blood vessels penetrating into fat; they are generally absent from the blood vessels of other tissues.[37] This suggests some sort of mutual interaction between fat cells and blood vessel stem cells that ensures that a ready supply of mitotic cells is continuously present within fat.

Finally, stem cells for more specialized structures also appear to exist within the body. For example, unspecialized cells residing within the pulp cavity of a tooth appear to have the ability to differentiate into odontoblasts and have the potential to form the entire root of a tooth.[11] This provides the hope that in the future, damaged teeth can be replaced with tooth buds derived from tooth stem cells. There is evidence that the bone marrow may harbor cells that can migrate to muscle, liver, cartilage, and pancreas and which may help repair damage to these organs.[30] A number of studies have also reported the existence of stem cells in blood and bone marrow that can migrate to the ovaries and replace degenerated egg cells.[38] If this is the case, reproductive senescence due to depletion of ovarian oocytes may be correctable. In summary, adult stem cells may have far more potential to effect repairs to the body than was dreamed of only twenty years ago.

References

1. Borisov AB, Raeker MO, Russell MW (2008) Developmental expression and differential cellular localization of obscurin and obscurin-associated kinase in cardiac muscle cells. *J Cellular Biochem* 103: 1621–1635.
2. Charge SPB, Rudnicki MA (2004) Cellular and molecular regulation of muscle regeneration. *Physiol Rev* 84: 209–238.
3. Charo IF, Taubman MB (2004) Chemokines in the pathogenesis of vascular disease. *Circulation Res* 95: 858–866.
4. Dawe HR, Farr H, Gull K (2007) Centriole/basal body morphogenesis and migration during ciliogenesis in animal cells. *J Cell Sci* 120: 7–15.

5. Di Padova M (2007) MyoD acetylation influences temporal patterns of skeletal muscle gene expression. *J Biol Chem* 282: 37650–37660.

6. Fox A, *et al.* (2007) Mobilization of endothelial progenitor cells into the circulation in burned patients. *Br J Surg* 95: 244–251.

7. Garry DJ, Olson EN (2006) A common progenitor at the heart of development. *Cell* 127: 1101–1104.

8. Hayashi S, Inoue A (2007) Cardiomyocytes re-enter the cell cycle and contribute to heart development after differentiation from cardiac progenitors expressing Isl1 in chick embryo. *Develop Growth Differ* 49: 229–239.

9. Hoenig MR, Bianchi C, Sellke FW (2008) Hypoxia Inducible Factor-1α, endothelial progenitor cells, monocytes, cardiovascular risk, wound healing, cobalt and hydralazine: A unifying hypothesis. *Current Drug Targets* 9: 422–435.

10. Hristov M, Wever C (2004) Endothelial progenitor cells: characterization, pathophysiology, and possible clinical relevance. *J Cell Mol Med* 8: 498–508.

11. Huang GT, *et al.* (2008) The hidden treasure in apical papilla: the potential role in pulp/dentin regeneration and bioroot engineering. *J Endod* 34: 645–651.

12. Jung J, *et al.* (2005) Jumonji regulates cardiomyocyte proliferation via interaction with retinoblastoma protein. *J Biol Chem.* 280: 30916–30923.

13. Kawai-Kowase K, Owens GK (2007) Multiple repressor pathways contribute to phenotypic switching of vascular smooth muscle cells. *Am J Physiol Cell Physiol* 292: C59–C69.

14. Kokovay E, Shen Q, Temple S (2008) The incredible elastic brain: how neural stem cells expand our minds. *Neuron* 60: 420–428.

15. Koster MI, Roop DR (2007) Mechanisms regulation epithelial stratification. *Annu Rev Cell Dev Biol* 23: 93–113.

16. Kucia M, *et al.* (2004) Cells expressing early cardiac markers reside in the bone marrow and are mobilized into the peripheral blood after myocardial infarction. *Circ Res* 95: 1191–1199.

17. Lafuste P, *et al.* (2005) ADAM12 and α9β1 integrin are instrumental in human myogenic cell differentiation. *Molec Biol Cell* 16: 861–870.

18. LeSimple P, *et al.* (2007) Trefoil factor family 3 peptide promotes human airway epithelial ciliated cell differentiation. *Am J Respir Cell Mol Biol* 36: 296–303.

19. Li H, *et al.* (2008) Research of PDGF-BB gel on the wound healing of diabetic rats and its pharmacodynamics. *J Surg Res* 145: 41–48.
20. Littlefield RS, Fowler VM (2008) Thin filament length regulation in striated muscle sarcomeres: Pointed-end dynamics go beyond a nebulin ruler. *Semin Cell Dev Biol* 19: 511–519.
21. Lu, H, *et al.* (2007) Combinatorial protein therapy of angiogenic and arteriogenic factors remarkably improves collaterogenesis and cardiac function in pigs. *Proc Natl Acad Sci USA* 104: 12140–12145.
22. Maguire EA, *et al.* (2003) Navigation expertise and the human hippocampus: a structural brain imaging analysis. *Hippocampus* 13: 250–259.
23. Mita S, Schon EA, Herbert J (1989) Widespread expression of amyloid beta-protein precursor gene in rat brain. *Am J Pathol* 134: 1253–1261.
24. Moniaux N, *et al.* (2001) Structural organization and classification of human mucin genes. *Front Biosci* 6: D1192–D1206.
25. Olguin HC, *et al.* (2007) Reciprocal inhibition between Pax7 and muscle regulatory factors modulates myogenic cell fate determination. *J Cell Biol* 177: 769–779.
26. Ordy JM, *et al.* (1993) Selective vulnerability and early progression of hippocampal CA1 pyramidal cell degeneration and GFAP-positive astrocyte reactivity in the rat four-vessel occlusion model of transient global ischemia. *Exp Neurol* 119: 128–139.
27. Pajalunga D, *et al.* (2007) Critical requirement for cell cycle inhibitors in sustaining nonproliferative states. *J Cell Biol* 176: 807–818.
28. Porte Y, Buhot MC, Mons NE (2008) Spatial memory in the Morris water maze and activation of cyclic AMP response element-binding (CREB) protein within the mouse hippocampus. *Learning & Memory* 15: 885–894.
29. Puzzo D, *et al.* (2008) Picomolar Amyloid-β positively modulates synaptic plasticity and memory in hippocampus. *J Neuroscience* 28: 14537–14545.
30. Ratajczak MZ, *et al.* (2008) Very small embryonic-like (VSEL) stem cells in adult organs and their potential role in rejuvenation of tissues and longevity. *Exp Gerontol* 43: 1009–1017.
31. Rota M, *et al.* (2008) Local activation or implantation of cardiac progenitor cells rescues scarred infarcted myocardium improving cardiac function. *Circ Res* 103: 107–116.

32. Senoo M, Pinto F, Crum CP, McKeon F (2007) p63 is essential for the proliferative potential of stem cells in stratified epithelia. *Cell* 129: 523–536.

33. Shawi M, Autexier C (2008) Telomerase, senescence and ageing. *Mech Ageing Dev* 129: 3–10.

34. Shireman PK, *et al.* (2007) MCP-1 deficiency causes altered inflammation with impaired skeletal muscle regeneration. *J Leukocyte Biol* 81: 775–785.

35. Shroyer NF, *et al.* (2005) Gfi1 functions downstream of Math1 to control intestinal secretory cell subtype allocation and differentiation. *Genes & Development* 19: 2412–2417.

36. Takahashi K, *et al.* (2007) Induction of pluripotent stem cells from adult human fibroblasts by defined factors. *Cell* 131: 861–872.

37. Tang W, *et al.* (2008) White fat progenitor cells reside in the adipose vasculature. *Science* 322: 583–586.

38. Tilly JL, Niikura Y, Rueda BR (2009) The current status of evidence for and against postnatal oogenesis in mammals: a case of ovarian optimism versus pessimism? *Biol Reprod* 80: 2–12.

39. Tskhovrebova L, Trinick J (2008) Giant proteins: sensing tension with titin kinase. *Current Biology* 18: R1141–R1142.

40. Wang, Y, *et al.* (2006) Evidence for ischemia induced host-derived bone marrow cell mobilization into cardiac allografts. *J Molec Cell Cardiol* 41: 478–487.

41. Whitwell JL, *et al.* (2007) 3D maps from multiple MRI illustrate changing atrophy patterns as subjects progress from mild congnitive impairment to Alzheimer's disease. *Brain* 130: 1777–1786.

42. Wu H, Zhou Y, Xiong Z-Q (2007) Transducer of regulated CREB and late phase long-term synaptic potentiation. *FEBS J* 274: 3218–3223.

43. You Y, *et al.* (2004) Role of f-box factor foxj1 in differentiation of ciliated airway epithelial cells. *Am J Physiol Lung Cell Mol Physiol* 286: L650–L657.

Chapter 4

HOW DO GIANT CELLS FORM?

With the exception of skeletal muscle cells, most cells in the body retain many of the aspects of size and shape that were possessed by their predecessors, the pluripotent stem cells of the embryo. To be sure, cells do acquire specialized proteins and organelle modifications to carry out their functions, but most do not radically change to acquire huge cellular proportions, multiple nuclei, or abnormal numbers of chromosomes. This relative uniformity of cells throughout the body makes the exceptions to these rules even more interesting. A number of cells in the body acquire relatively gigantic proportions, and each of these cell types accomplishes this through different mechanisms. By studying aberrant cells with enlarged nuclei or size, it is easier to understand the mechanisms that limit overall size in normal cells. This chapter will showcase these aberrant cells.

1. Megakaryocytes and the Regulation of Chromosome Number

Megakaryocytes constitute about 0.06% of the nucleated cells found within the hollow cavities within bones (bone marrow)[27] (Fig. 4.1). Each cell has a diameter 5–10 times the diameter of other bone marrow cells and has an enlarged, highly folded nucleus. Megakaryocyte nuclei are polyploid, that is, they usually contain eight times the normal amount of DNA (16N); some cells can contain sixty-four times the normal amount (128N)![48]

In addition to having abnormal nuclei, megakaryocytes have very peculiar, enlarged cytoplasms. Each cell settles down upon thin-walled, venous sinusoids of the bone marrow and then extrudes long, thin cytoplasmic processes into the lumen of the sinusoid. These

81

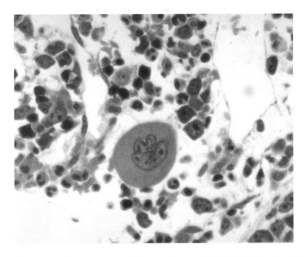

Fig. 4.1.   A large megakaryocyte with an enlarged, highly folded nucleus. This cell is surrounded by immature blood cell precursors and by thin walled venous sinusoids.

processes arise from compartments of cytoplasm called platelet demarcation zones. Each platelet demarcation zone, in turn, is created when tubular infoldings of the plasma membrane break off from the cell surface and then merge to form large, flattened, membrane-bound bubbles that surround a localized area of cytoplasm. Eventually, each tubular cell process fragments into 6–7 teardrop-shaped bits of cytoplasm called platelets that are carried away into the bloodstream. This process rather resembles the tearing off of tickets from a roll of ticket paper in a carnival. By producing many such fragmenting processes, a megakaryocyte can produce as many as 3000 platelets before undergoing cell death.[22] The huge size of the megakaryocyte cytoplasm is helpful in producing as many platelets as possible. Also, platelets from polyploid megakaryocytes appear to be larger and more reactive than those from megakaryocytes that do not achieve polyploidy.[25,35,36]

The function of platelets is well known: they accumulate at breaks in blood vessels and help form a clot, as well as secreting growth factors that stimulate wound healing. But what causes their formation and what is responsible for the monstrous anatomy of their parent cells, the megakaryocytes?

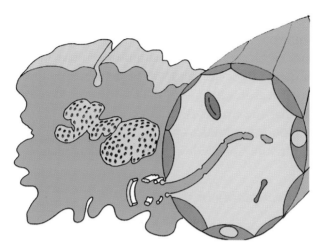

Fig. 4.2. Diagram of a megakaryocyte, showing how cell processes become fragmented into blood-borne platelets.

Megakaryocyte precursor cells of the bone marrow first appear to develop under the influence of several growth factors (interleukin-3, stem cell factor) produced by the stromal cells of bone marrow.[7] However, they do not completely differentiate unless they are exposed to a hormone called *thrombopoietin*, which was discovered in 1994.[32] Almost all of the body's *thrombopoietin* (95%) is continuously produced by hepatocytes of the liver and is carried to the bone marrow to stimulate platelet production.[46] Platelets and megakaryocyte precursors bind *thrombopoietin*, respond to it, and then internalize and destroy this hormone. So, if numbers of platelets in the blood are sufficient, *thrombopoietin* will be continuously degraded and blood levels of it will be modest. If platelet production falls, it will not be degraded and blood levels of *thrombopoietin* will rise to stimulate the birth of more megakaryocytes.

At first glance, it would seem surprising that the liver would regulate the function of a seemingly unrelated tissue, the bone marrow. However, there are two reasons why this actually makes sense. First, the liver in the embryo is a major site of blood cell formation. It is only after the replacement of the limb cartilages by bone that blood-forming cells abandon the liver and take up residence within the marrow cavities of long bones. Why do they do this? The predominance of blood formation within bone marrow is also part of the story of megakaryocytes, and will be explained below.

The second reason why it is logical for the liver to regulate platelet formation is that the liver is the source of most of the clotting proteins in the blood that bind to platelets and help form clots. So it is appropriate, after all, for the liver to also regulate platelet formation.

Induction of polyploidy in megakaryocytes

How does thrombopoietin induce polyploidy in megakaryocyte precursor cells? One recent study has shown that mature megakaryocytes fail to produce a specific protein called *Polo-like Kinase* (*PLK*). Also, if megakaryocytes are artificially forced to express *PLK*, they fail to become polyploid.[48] Clearly, *PLK* has an important role to play in the regulation of mitosis and *polyploidy*. What is *PLK* and how does it exert its effects in megakaryocytes?

PLK is the mammalian analogue of a protein first discovered in Drosophila by Christian Nüsslein-Volhard, who also was a pioneering investigator of homeotic proteins. *PLK* was first characterized as a mutated protein that caused abnormalities in the poles of the mitotic spindle, and so was called *polo*. It is now known that *PLK* is an important regulator of proteins that control mitosis.[24]

One aspect of mitosis that *PLK* regulates is the adhesion of chromosomes to one another. After the S phase of a cell and the duplication of DNA, a single chromosome is transformed into 2 identical copies of DNA; each copy is then termed a chromatid. Later, during mitotic prophase, each chromatid becomes greatly condensed, but the two sister chromatids are tightly bound to each other along their entire lengths by a DNA-binding protein called *cohesin*.

PLK helps release the chromatids from one another by phosphorylating and inactivating *cohesin*, which then detaches from the chromatids (Fig. 4.3). However, at one region of the chromatid, the centromere, *cohesin* molecules are protected from the action of *PLK* by a protein called *Sgo1* and its associated phosphorylase enzyme, so that the

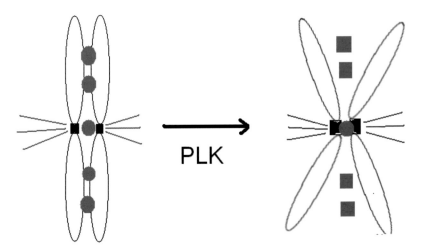

Fig. 4.3. Diagram of the transition between prophase of mitosis (left) and metaphase. In prophase, sister chromatids are bound together by *cohesin* (red circles). After exposure to *polo-like kinase* (*PLK*), *cohesin* becomes phosphorylated (red squares) and dissociates from chromatids during metaphase.

chromatids remain partially attached to one another during metaphase.[31] How is this partial attachment, then, broken to permit anaphase?

An answer to this question requires a look at the kinetochore, the protein structure that links the centromeres to microtubules. Each kinetochore is a very complicated structure, composed of at least 80 proteins that form about 10 protein complexes; each one of these complexes has a name formed from acronyms that arose during the discovery of each complex (Fig. 4.4).

At one end of each kinetochore, the *Dam1* protein complex forms a ring around a cluster of microtubules and acts like an engine that pulls the entire structure along microtubules towards the mitotic centrioles. At the other end of a kinetochore, *CENP* complexes bind to the chromatin (DNA plus associated proteins like histones) of the chromatids. The region of the chromosome that binds the kinetochore (the centromere) contains long sequences of non-coding DNA that are repeated many times; kinetochore proteins apparently recognize these specific DNA sequences and thus localize to the correct region of the chromosome.

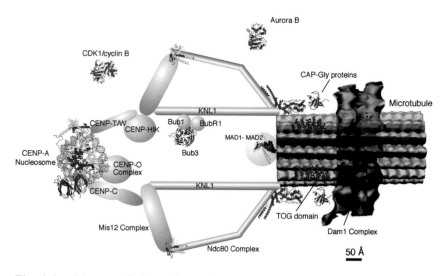

Fig. 4.4.   Diagram of a kinetochore, showing 8 complexes of proteins with specific functions.

Other structural protein complexes of the kinetochore have other functions. For example, the *MAD* complex seems to somehow sense whether or not microtubules are correctly anchored to the kinetochore, whereas the *Bub* complex senses how much tension is being exerted upon the kinetochore.[38,44]

When all the chromatids of a cell are firmly attached to microtubules and under tension, the proteins of the kinetochore somehow send a signal that initiates anaphase. This signal is sent to another complex of 13 proteins that binds to the kinetochore; this additional complex is called the *Anaphase Promoting Complex (APC)*.[16,42]

The *APC* is a critical regulator of anaphase. It works by attaching molecules of *ubiquitin* to proteins. *Ubiquitin* acts like a "black flag" upon proteins and attracts the attention of destructive, miniature machines called proteosomes. Any protein tagged with *ubiquitin* is shortly destroyed within proteosomes, which function rather like tiny, intracellular meat grinders or wood chippers. *Ubiquitin*-tagged proteins are fed into one end of each barrel-shaped proteasome, and only peptide fragments come out the other end. The discoverers of this destructive system, which is important for the regulation of all cellular proteins, were awarded the Nobel Prize in 2004.[43]

What proteins are targeted for destruction by the *APC*? One protein is a blocking protein called *securin*. *Securin* normally protects the *cohesin* holding together the chromatids at the centromere. When *securin* is destroyed by the *APC*, this allows the action of an enzyme called *separase*, which acts to destroy the *cohesin* remaining on the chromatids and finally allows the complete separation of the chromatids during anaphase (Fig. 4.5).

How does this all relate to our problem of understanding megakaryocyte polyploidy and the effects of *PLK*? Most investigators think that megakaryocytes fail to complete anaphase, so that duplicated chromosomes remain attached to each other and come to reside within an enlarged nucleus.[32] A failure of *PLK* to act could explain this, because *PLK* may also phosphorylate and activate the *APC*.[38] Thus, the lack of *PLK* seen in megakaryocytes could inactivate the *APC* and may explain both an adhesion of prophase chromatids to each other and an abnormal anaphase in these cells.

The *APC* not only causes the destruction of *securin*, but also ubiquitinates and destroys *cyclin B* and its *cyclin-dependent kinase*.

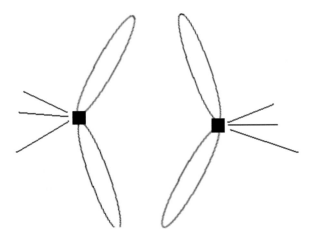

Fig. 4.5. Diagram of anaphase, showing how sister chromatids are pulled apart from each other by traveling along spindle microtubules.

This permits the completion of mitosis. So, if *PLK* does not work properly, this dysfunction could explain many aspects of abnormal mitosis in megakaryocytes.

One question that all of the above does not answer is this: why are additional rounds of DNA synthesis permitted in megakaryocytes? Most cells undergo DNA synthesis only one time per cell cycle and only after successful completion of mitosis. These rules must be broken in megakaryocytes to allow the accumulation of many copies of each chromosome in the nucleus. What are the rules of DNA synthesis, and how are they violated in megakaryocytes?

DNA synthesis in cells is potentially a very laborious procedure. Each cell contains strands of DNA that total almost a meter in length. Even when broken into 46 chromosomes, this tremendous amount of DNA would take a very long time to duplicate if *DNA polymerase* began its copying function at the end of each chromosome. Instead, chromosomal DNA duplication is simultaneously initiated at about 25,000 sites called replication origins that are located at regular intervals along DNA.[4] At each site, a so-called *origin recognition complex* of proteins binds to DNA and recruits other proteins (*Cdc6, Cdt1,* and the *minichromosome maintenance complex*) to join in the action.[26,29] All of these proteins recruit the DNA-copying enzymes (*DNA polymerases*) to the replication origin to begin DNA copying. In addition, the *minichromosome maintainance complex* functions as a *helicase* and partially unwinds the DNA double helix to permit the enzymes to have access to the DNA. However, none of these proteins actually start DNA synthesis unless they are activated by *Cyclin E* and its *cyclin-dependent kinase.*

Most mammalian cells have a safeguard protein that prevents DNA synthesis from occurring more than once prior to mitosis. This protein, called *geminin,* causes the *minichromosome maintenance proteins* to fall off of the DNA after the S phase, so that excess DNA copying cannot occur. However, once mitosis is safely completed, the *Anaphase Promoting Complex* destroys *geminin.* This allows for another round of DNA synthesis (S phase), which will be ended once again by the reappearance of more *geminin.*[26,29] The carefully regulated activation or destruction of all of these proteins underlies the regular transitions of cells from the S phase of DNA synthesis to the M phase of mitosis.

It is still not known with certainty how all of these proteins relate to polyploidy in megakaryocytes. It is interesting, however, that of 100 genes that are down-regulated during megakaryocyte development, 24 of them belong to the category of genes that regulate DNA replication, including the *minichromosome maintenance complex.*[35]

Megakaryocyte cytoplasm

Reasons why the huge cytoplasm of megakaryocytes develops are also worth considering. Most cells maintain a relatively rigid ratio of nucleus to cytoplasm. Moreover, cells often enter mitosis only after the overall volume of the cell has grown large enough to require or support it. So, there is a link between cell division and cell size that seems abnormal in megakaryocytes. What is this link?

One important link between cell size, cell nutrition, and the control of mitosis is a kinase enzyme called *TOR*. *TOR* stands for *Target of Rapamycin*. This protein was originally identified in yeast as an enzyme inhibited by a bacterial antibiotic called rapamycin. Rapamycin appears to mimic a state of starvation in cells and causes cells to limit their growth and overall size. Rapamycin does this by inhibiting the *TOR* protein, an action which prevents the cell from synthesizing any more proteins or from entering mitosis. One way the *TOR* protein does this is by phosphorylating (activating) a protein called S6 kinase, which in turn regulates a component of ribosomes (*S6* protein). When the *S6* protein is not activated by *TOR*, the proteins required for ribosome assembly and function are no longer translated and produced, and there is a generalized fall in the production of ribosomes. Without ribosomes, fewer proteins can be produced to replace those degraded by the *ubiquitin* pathway and proteasomes. In short, inhibition of *TOR* puts a cell into sort of a state of hibernation until nutrient supplies improve.[28] Changes in the *TOR* pathway are likely reasons why the cytoplasm of a megakaryocyte is allowed to grow so large.[27]

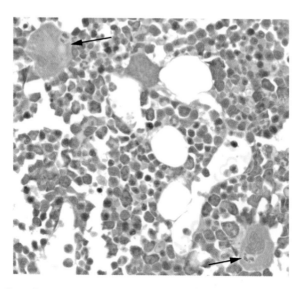

Fig. 4.6.   Overall view of bone marrow, showing two megakaryocytes that have engulfed neutrophil cells within their cytoplasms (arrows).

Another very strange aspect of megakaryocyte cytoplasm is that about 5% of megakaryocytes can be "caught in the act" of actively engulfing other cells such as neutrophils or red blood cells and carrying them about as the move through the bone marrow[8] (Fig. 4.6)! The significance of this peculiar activity of swallowing other cells, called emperipolesis, is not clear. In normal animals, the engulfed cells appear normal and may be released following their entrapment. This process of cell engulfment is not unique to megakaryocytes and also occurs within cells of the thymus, where it may play a role in the development of lymphocytes (see Ch. 5). In genetically modified animals that lack a megakaryocyte transcription factor called *GATA-1*, megakaryocytes fail to mature properly and become much more numerous; in addition, the fraction of megakaryocytes that engulf cells increases to 34%!

Another abnormality that develops in mice lacking the *GATA-1* gene is that they develop massively thickened bones and a diminished size of bone marrow cavities.[19] This occurs because the numerous, immature megakaryocytes of such animals exhibit an enhanced potential to stimulate bone-forming cells called osteoblasts. The mechanism whereby megakaryocytes stimulate bone formation has not been precisely identified, but seems to require direct contact between megakaryocytes and osteoblasts.[11]

Osteoblasts themselves seem to be able to stimulate the formation of megakaryocytes and other hematopoietic (blood forming) cells.[19] Also, osteoblasts secrete *stromal cell derived factor-1*, which causes hematopoietic cells to migrate into bone marrow.[41] Thus, a reciprocal interaction between megakaryocytes and bone cells ensures that most blood cell formation in the adult occurs within the protective armor of the bones. This is not only helpful for the protection of critical blood cells, but also serves to provide a protected environment for other vital stem cells of other tissues (endothelia, cardiac muscle, etc., described in Ch. 3). Thus, our increasingly clearer understanding of the functions of these specialized, giant megakaryocytes has given us an unexpected insight into the wisdom of nature that has arranged for a protective niche for stem cells.

2. Osteoclasts, Giant Cells of Bone

Another type of giant cell is often found quite close to megakaryocytes, but has an entirely different anatomy and function. These cells, called osteoclasts, are huge cells found upon the surfaces of bone, either within the marrow cavity or external to it (Fig. 4.7). Osteoclasts commonly contain between 10–20 cell nuclei, although a few can contain as many as 40 nuclei.[6] These cells have a critical influence upon bone turnover and bone healing. This topic in general is of considerable practical interest: out of the 6 million fractures occurring annually in the United States, about 500,000 cases show delayed or impaired healing.[13] So a better understanding of the remodeling of bone by osteoclasts has great medical potential.

The function of osteoclasts involves adhering to specific sites in bone and secreting two types of molecules: they secrete *lysosomal cysteine proteases*, enzymes that function well in an acid environment and which can degrade bone proteins (*Cathepsin K* is the major one), and they also secrete acid, which can dissolve the mineral component

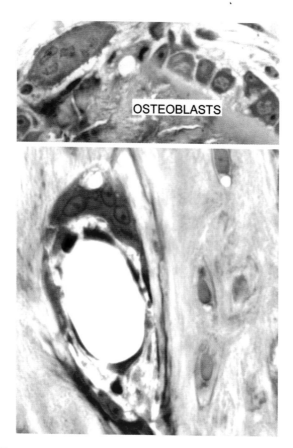

Fig. 4.7. Two views of multinucleated osteoclasts, adjacent to bone-forming osteoblasts or within a hollowed-out space within bone. Only a few of the many nuclei of each cell are visible in these single sections through these large cells. In the lower picture, an osteoclast has excavated a space within bone and is surrounded by bone matrix and by osteocytes that live in small spaces (*lacunae*) in the bone matrix.

of bone.[17] Where do these cells come from, and what causes them to erode bone only where needed and leave healthy bone intact?

The progenitors of osteoclasts are neighboring cells of the bone marrow called monocytes. Monocytes arise from bone marrow stem cells that have the potential to form either neutrophils or monocytes. When these stem cells are exposed to a specific bone-marrow derived growth factor (*granulocyte-macrophage colony stimulating factor*, or

GM-CSF), they differentiate into monocytes and generally leave the bone marrow and enter the bloodstream. Monocytes often leave the blood to enter ordinary connective tissue and become phagocytic cells called macrophages. When necessary, however, blood vessel cells in bone marrow can secrete a homing protein called *stromal cell derived factor-1* that stimulates monocytes to travel back to the bone marrow and differentiate into osteoclasts.[41,47]

Turning on the genetic program for monocytes by these growth factors seems to involve the activation of a homeotic protein called *HOX-A10* and the promotion of the transcription of mRNAs specific for monocytes.[40] When monocytes gather around an area of defective bone, they are somehow stimulated to fuse together to form giant osteoclasts. This bizarre abandonment of individual cell identity seems to involve a number of membrane proteins that promote membrane fusion. One such protein is called *DC-STAMP* (it was originally found on *Dendritic Cells*, descendents of mono-cytes/macrophages that inhabit lymphoid organs).[49] Other membrane proteins involved in this process are called *tetraspanins*, which belong to a family of 20 related proteins that all have four membrane-spanning domains.[18]

Why is cell fusion necessary? Each osteoclast forms a large, suction-cup like structure that adheres to bone and dissolves the bone beneath it. In order to maintain the large cytoplasm of the osteoclast, it would seem to make sense to either increase the number of chro-mosomes within one nucleus or increase cell chromosomes by increasing the number of nuclei. Osteoclasts have chosen the second option. Since studies have shown that the ability of an osteoclast to resorb bone increases linearly with the numbers of its nuclei, cell fusion appears to be a functional necessity for these cells.[6]

What controls the activity of osteoclasts? One protein produced by bone-forming osteoblasts, termed *RANK-ligand*, binds to osteo-clasts and stimulates their differentiation and development. *RANK-ligand*, for example, induces the appearance of *tetraspanins* in osteoclasts that are needed for cell fusion. Hormones called *calcitonin* and *parathyroid hormone* may also directly regulate osteoclast activity, though the existence of parathyroid hormone receptors on osteoclasts is still the subject of some debate.[12] Perhaps a more specific, and more important, question is why osteoclasts avoid some areas of bone and destroy others?

Signals from other bone cells seem to be important for this selection process. For example, bone-forming osteoblast cells can partly remove the incompletely mineralized bone matrix beneath them by secreting an enzyme called *collagenase.* This allows the osteoclasts to attach themselves to hard, mineralized bone and begin bone degradation.[17]

Another mechanism appears to involve the detection of abnormalities in the environment of the *osteocytes* embedded in bone. Each osteocyte inhabits a tiny, fluid-filled chamber called a lacuna, and each lacuna is connected to adjacent ones via thin channels, called canaliculi, that form tiny tunnels in the surrounding bone matrix. All the osteocytes possess a sensory organ called a primary cilium, which senses changes in the fluid flowing through the thousands of canaliculi within bone. When this fluid flow becomes abnormal, as may occur after a small fracture, the osteocytes release *prostaglandins* into their environment, which stimulates the activity of osteoclasts and the remodeling of bone in the abnormal area.[45]

3. Adipocytes, Giant Cells of Connective Tissue

Most of the lipid in the body is stored within greatly enlarged cells (adipocytes) of so-called white fat (Fig. 4.8). These cells can also acquire diameters of 100 microns or more, and thus are as large or larger than megakaryocytes. Unlike megakaryocytes, fat cells have modest numbers of intracellular organelles because the huge cytoplasm of a fat cell is mainly occupied by an enormous lipid droplet. Thus, the ordinary-sized, flattened nucleus of a fat cell apparently contains enough genetic material to maintain cell processes, and so need not be polyploid or duplicated. Fat cells do, however, face their own unique challenges.

A major function of fat cells is to remove triglycerides (or triacylglycerols) from the circulation for storage. This poses a problem. A triglyceride is composed of three long molecules of fatty acids that

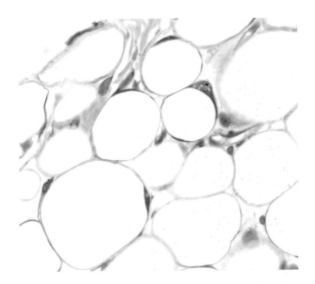

Fig. 4.8. Some examples of adipocytes. Lipids in fat cells are largely extracted from the cell by organic solvents during preparation for histology, leaving behind large, pale-staining spaces that were once filled with fat.

are joined to a small "backbone" of glycerol. Each huge molecule is thus mainly hydrophobic, so that it is not soluble in water and must be carried through the bloodstream in association with a large carrier protein (*albumin*). A triglyceride molecule is simply too big to pass into a fat cell through pores in the plasma membrane: a suitable pore would allow the cytoplasm itself to leak out. So how do fat cells accumulate fat?

Adipocytes partly solve this problem by synthesizing an enzyme, *lipoprotein lipase*, which can break down triglycerides into smaller fatty acids and glycerol. However, this action is not performed at the surface of the fat cell, but within the lumena of nearby capillaries. The *lipoprotein lipase* protein is exported from the fat cell, binds to carbohydrate-rich proteoglycan molecules on the outer surface of a capillary, and then is internalized and dragged through the cytoplasm of the endothelial cell until it is exposed upon the inner surface of the capillary.[30] Triglycerides passing nearby in the bloodstream are thus broken into smaller components, which can pass through the capillary wall and reach fat cells. Fat cells are not the only cells that do this: muscle cells that mainly rely upon the burning of triglycerides as an energy source also utilize this action of *lipoprotein lipase.*

How are the smaller components of fat internalized? Glycerol is a small, polar molecule that is easily taken up into fat cells via a transporter protein that forms small pores in the plasma membrane. This transporter protein is called *adipocyte aquaporin,* and belongs to a larger family of 10 related proteins that also help transport water across the lipid bilayers of cell membranes.[20] *Aquaporins* are so-called *multipass membrane proteins* that are threaded into and out of the plasma membrane and which form pores in the membrane. When fat cells are stimulated by *norepinephrine,* membranous vesicles enriched in *aquaporin* are stimulated to migrate towards the cell membrane. This adds *aquaporin* transporters to the cell membrane and makes it much more permeable to glycerol. Movement of vesicles bearing transporter proteins to the cell membrane is a general method by which cells regulate their permeability: transporters of glycerol, water, or glucose can all be added or subtracted to the membrane to change the uptake of small molecules.[20]

Fatty acids pose a more substantial problem: how are these fat-soluble molecules removed from the lipid-rich plasma membrane and taken up into the cytoplasm? A number of fat-cell specific proteins, such as *fatty acid binding protein* and *fatty acid transport protein,* seem required for this process, but the details are still murky.[33]

Where do fat cells come from? They arise from precursor cells that resemble fibroblasts.[15] As they differentiate, under the influence of specific types of *Bone Morphogenetic Protein* (remember the importance of this protein, discussed in Ch. 2, for the induction of mesodermal cell fates?), they begin to express two types of transcription factors that stimulate the expression of fat-cell specific genes. One type of transcription factor belongs to the *peroxisome-proliferator-activated receptor* family of proteins (*PPAR*); the other type belongs to the *C/EBP* family, named after the ability of these proteins to bind to an enhancer gene sequence called CCAAT (*C Enhancer Binding Protein*). These transcription factors seem sufficient to trigger all the developmental changes that transform fibroblasts into fat cells.

A disruption of the actions of these transcription factors can have dramatic results. For example, if *C/EBP* is blocked experimentally in developing mice, the mice are born with virtually no fat cells! This may seem like a desirable outcome in the context of the current epidemic of obesity, but in fact it has disastrous consequences for these mice. When the mice are young and still ingesting maternal milk, they always have a rapidly digestible source of energy right within their stomachs and so have a normal metabolism. However, when they are weaned and eat more slowly digestible solid food, they have no way to store energy-rich molecules to power their muscles and body, because they have no fat cells. Thus, when fasted, they have no immediate access to sources of emergency calories. Such mice rapidly use up the glycogen stored in the liver after a meal and then "run out of gas." Muscles cannot use lipids released from fat cells for energy and so become inert. The brain, which usually depends upon the breakdown of glycogen into glucose for energy, also falters. Such mice sink into a severe torpor, become inactive, and have body temperatures that plummet to close to air temperature.[14] These defects illustrate how useful it is for the body to have a reservoir of energy-rich fat that can be called upon in time of need or added to during times of feasting.

Another function of fat cells is to produce a protein hormone called *leptin*. *Leptin* is carried to a specific part of the brain, the arcuate nucleus of the hypothalamus, to signal the brain that fat depots are increasing or decreasing, and is part of a major mechanism that regulates overall body fat.[50] Genetically obese strains of mice, such as the *db/db mouse*, lack functional receptors for *leptin* in the hypothalamus. This causes the hypothalamus to react as if the body were starving and devoid of fat, so that animals overeat in spite of a normal food supply. Such *leptin*-insensitive animals become grossly obese and diabetic (Fig. 4.9).

Do defective *leptin* receptors cause obesity in humans? While rare cases of genetic defects in the *leptin* signaling system have been found in a few human families, it is clear that human obesity is overwhelmingly **not** due to mutations in *leptin* or its receptors. It is known that chronic consumption of calorie-rich diets can somehow cause the hypothalamus to become resistant to *leptin* by mechanisms that are

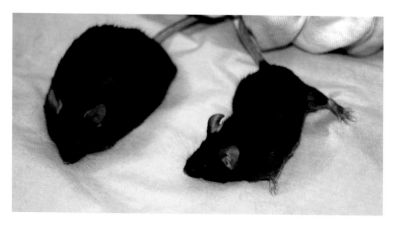

Fig. 4.9. A normal mouse (right) and a genetically obese mouse (left) that lacks functional *leptin* receptors in the hypothalamus.

not completely understood. A better understanding of the neuroendocrine control of body weight by *leptin* could potentially have important implications for the control of obesity and obesity-related diabetes.

Another aspect of fat cells that is of interest is that there is a well-known sexual dimorphism in fat deposition: women have higher levels of body fat than men, and also have sexually dimorphic patterns of fat deposition on the body. Women tend to accumulate fat on the hips and breasts, whereas men tend to develop increases in intraabdominal and belly fat. These sex differences are partly due to effects of sex steroids on adipocyte-specific proteins. *Estrogen* alters the levels of some types of *C/EPB proteins* in fat cells; curiously, these effects of *estrogen* are different in subcutaneous fat cells relative to effects in intraabdominal fat cells.[21] Why *estrogen* and its *estrogen* receptors have different effects in different cells is one of the unsolved problems in endocrinology. *Estrogen* also has potent effects on a protein called *perilipin*, which surrounds the lipid droplet of fat cells and which regulates how much lipid is withdrawn from fat depots.[2]

One final problem for fat cells that needs discussion is how they acquire their final dimensions. This is not simply a matter of importing lots of lipid until the cell swells to enormous proportions.

The precursors of fat cells are fibroblasts, and like most fibroblasts, they are surrounded by masses of extracellular fibers composed of the *collagen* that they secrete. In order to expand in size, pre-adipocytes must secrete *matrix metalloproteinases*, enzymes that use zinc as a co-factor. These enzymes dissolve the *collagen* around the pre-adipocytes. If the enzymes are experimentally blocked, growing fat cells become strangled in a meshwork of extracellular fibers and never acquire the size that they should attain.[10]

4. Oocytes, Giant Cells of the Ovaries

Female germ cells, or oocytes, are some of the largest cells in the body; they may attain diameters 10-fold larger than most cells and possess a huge nucleus and a nucleolus that may be as large as the entire nucleus of a normal cell (see Fig. 2.4 for a reminder of the anatomy of these cells). These cells are neither polyploid nor multinucleate, but still manage to somehow maintain the functions of a very large cytoplasm. Where do these cells come from, and what makes them so large?

The precursors of oocytes, primordial germ cells, do not even originate within the ovary. Early in embryogenesis, these cells become sequestered into a pool of cells located in the embryonic yolk sac. They remain there until the gonads begin to secrete a protein called *stromal-cell derived factor*. Primordial germ cells then migrate to the gonads by following a gradient of *stromal-cell derived factor* and then burrow into the substance of each gonad.[3] If the gonad is female, a primordial germ cell becomes an oocyte and induces the surrounding connective tissue cells to form a protective structure called an ovarian follicle around itself. This inductive ability of oocytes to form follicles can be demonstrated in several ways. If oocyte-specific homeotic genes are deleted from an oocyte, it remains healthy but fails to induce the formation of a follicle.[34] Also, occasionally, oocytes will "lose their way" to the gonads and implant into the adrenal glands instead. If this happens, ovarian follicles will also form in the adrenal glands!

When a primordial germ cell enters the ovary, it is also stimulated to begin an initial meiotic division that will eventually reduce the number of chromosomes present in a mature oocyte. If, however, a primordial germ cell enters a male gonad (presumptive testis) it continues to undergo normal mitotic divisions until much later in life, and will eventually mature into a sperm cell. What features of the developing gonads cause these primordial germ cells to choose between meiosis and mitosis and female and male lives?

Recent studies have shown that a molecule called retinoic acid, produced by the ovaries, forces primordial germ cells to become meiotic oocytes. This does not happen in the developing testis, because testicular cells produce an enzyme that degrades retinoic acid. But how does retinoic acid cause oocytes to accept a female fate?

Two molecules introduced to you in Chapter 2 — *nanos* and *DAZL* proteins — form the basis for an answer to this question. You will recall that *nanos* was first identified as a key protein that established an initial morphogenic gradient in a fruit fly egg. The pole of the egg marked by *nanos* will eventually produce the germ cells of flies. Also, *DAZL* was found to be essential for the generation of germ cells in frogs. We now know that these proteins are of critical importance in mammals, as well. If *DAZL* is deleted from mammalian oocytes, they become unable to respond to retinoic acid and will not initiate meiosis.[23] Also, *nanos* has been found to initiate the male-specific genetic program of sperm cells and renders them insensitive to retinoic acid.[39] So, these two proteins steer the primordial germ cells towards either a male- or female-specific developmental pathway (Fig. 4.10).

Exactly how *nanos* and *DAZL* proteins regulate the response to retinoic acid, and how retinoic acid itself works to regulate the genes for meiosis, are questions that have not been addressed yet. At least, however, the key steps in these two developmental pathways have been identified.

Oocytes are peculiar because, even though they initiate meiosis much earlier than spermatocytes, they halt the progression of meiosis at the first meiotic prophase and then become suspended in a strange version of the cell cycle known as diplotene. Diplotene in oocytes may persist for 30 years! Why does this happen? The follicle cells that surround the oocyte appear to produce a number of unknown factors that bind to receptors on the oocyte. These factors bind to a receptor

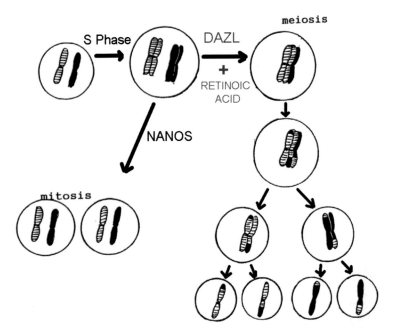

Fig. 4.10. Diagram of how a germ cell chooses between mitosis and meiosis. In mitosis, maternally derived chromosomes (black) and paternally derived chromosomes (grey) are duplicated and distributed equally between 2 daughter cells. In meiosis, only a paternally derived chromosome or a maternally derived chromosome is retained in each of 4 daughter cells.

called *GPR3* and stimulate the production of *cyclic AMP* in oocytes. This somehow maintains meiotic arrest.

This spell is broken by a surge in blood levels of pituitary hormones (*gonadotropins*) that occurs just prior to ovulation of the oocyte and release of the oocyte from the ovary. This surge of hormones stimulates the surrounding follicle cells to secrete an unusual variant of *insulin*, termed *insulin-like protein 3*, which was first discovered in the Leydig cells of the testis. This reproductive form of insulin binds to other receptors on oocytes and causes them to degrade cytoplasmic *cAMP*. This frees the cell from inhibition and allows the resumption of meiosis.[37]

The triumphant fate of a mature oocyte is, however, reserved for only a fortunate few. Out of the hundreds of thousands of oocytes present in the ovary of a young girl, only about 500 will eventually reach a state of maturity that allows them to be ovulated. The precise reasons why one oocyte is chosen over all of its neighbors remains one of the most puzzling and challenging questions for reproductive biology.

What causes the dramatic enlargement of the oocyte? A number of proteins produced by surrounding follicular cells, like *KIT ligand, fibroblast growth factor 7*, and *epidermal growth factor*, seem to be required for the stimulation of oocyte growth.[9] Also, effects of *growth hormone*, secreted by the pituitary, may be involved.[5] It makes functional sense for an oocyte to develop an enormous cytoplasm, since later, after fertilization, this cytoplasm must divide into smaller daughter cells that compose the embryo.

The cytoplasm of an oocyte is remarkable not only for its huge size, but also because of special properties not found in any other cell. When the nucleus of a sperm cell is introduced into the cytoplasm of an oocyte, cytoplasmic factors completely "re-program" the entire genome of that nucleus so that it can later express genes suitable for embryonic development and pluripotency. In fact, this amazing property of oocyte cytoplasm is central to the well-known practice of cloning, in which the nucleus of a skin cell is introduced into an enucleated egg cell and stimulated to divide. The result of this is not a pile of skin cells, as might be expected, but a normal, growing embryo with pluripotent embryonic stem cells. How is this dramatic transformation of an inserted nucleus carried out by oocyte cytoplasm?

The details of this magical event are not at all well understood. It is known that exposure to oocyte cytoplasm causes a number of things: changes in the acetylation or methylation of DNA-associated proteins called histones, a general decondensation of chromatin to make it accessible to enzymes, a stimulation of the genes required for pluripotency, and a suppression of normal somatic genes. One oocyte protein called *nucleoplasmin* has been implicated in all of this. Also, the very framework encompassing the nucleus undergoes changes. In normal somatic cells, a network of *lamin A* proteins interconnects all the nuclear pores on the interior of the nuclear envelope and helps shape the nucleus itself. After exposure to oocyte cytoplasm, the *lamin A* proteins are changed to a slightly different type (*lamin B*), which seems to permit a greater developmental plasticity of the cell.[1] While some of the factors that transform oocytes into giant cells are becoming better understood, it is clear from the above that the poorly characterized transformations and functions of oocytes make them the most mysterious of the giant cells of the body.

References

1. Alberio R, Campbell KH, Johnson AD (2006) Reprogramming somatic cells into stem cells. *Reproduction* 132: 709–720.
2. Akter MH, *et al.* (2008) Perilipin, a critical regulator of fat storage and breakdown, is a target gene of estrogen receptor-related receptor α. *Biochem Biophys Res Comm* 368: 563–568.
3. Ara T, *et al.* (2003) Impaired colonization of the gonads by primordial germ cells in mice lacking a chemokine, stromal cell-derived factor-1 (SDF-1). *PNAS USA* 100: 5319–5323.
4. Arias EE, Walter JC (2007) Strength in numbers: preventing re-replication via multiple mechanisms in eukaryotic cells. *Genes Dev* 21: 497–518.
5. Bevers MM, Izadyar F (2002) Role of growth hormone and growth hormone receptor in oocyte maturation. *Mol Cell Endo* 197: 173–178.
6. Boissy P, *et al.* (2002) Transcriptional activity of nuclei in multinucleated osteoclasts and its modulation by calcitonin. *Endocrinology* 143: 1913–1921.
7. Cardier JE, *et al.* (1996) Megakaryocytopoiesis *in vitro*: from the stem cells' perspective. *Stem Cells* 14(S.1): 163–172.
8. Centurione L, *et al.* (2004) Increased and pathologic emperiopolesis of neutrophils within megakaryocytes associated with marrow fibrosis in GATA-1[low] mice. *Blood* 104: 3573–3580.
9. Cho J-H, *et al.* (2008) Fibroblast growth factor 7 stimulates in vitro growth of oocytes originating from bovine early antral follicles. *Molec Reprod Devel* 75: 1736–1743.
10. Chun T-H, *et al.* (2006) A pericellular collagenase directs the 3-dimensional development of white adipose tissue. *Cell* 125: 577–591.
11. Ciovacco WA, *et al.* (2009) The role of gap junctions in megakaryocyte-mediated osteoblast proliferation and differentiation. *Bone* 44: 80–86.

12. Dempster DW, *et al.* (2005) Normal human osteoclasts formed from peripheral blood monocytes express PTH type 1 receptors and are stimulated by PTH in the absence of osteoblasts. *J Cell Biochem* 95: 139–148.

13. Dimitriou R, Tsiridis E, Giannoudis PV (2005) Current concepts of molecular aspects of bone healing. *Injury* 36: 1392–1404.

14. Gavrilova O, *et al.* (1999) Torpor in mice is induced by both leptin-dependent and -independent mechanisms. *Proc Natl Acad Sci USA* 96: 14623–14628.

15. Gesta S, Tseng Y-H, Kahn CR (2007) Developmental origin of fat: tracking obesity to its source. *Cell* 131: 242–250.

16. Han D, *et al.* (2009) Crystal structure of the N-terminal domain of anaphase-promoting complex subunit 7. *J Biol Chem.*

17. Holliday LS, *et al.* (1997) Initiation of osteoclasts bone resorption by interstitial collagenase. *J Biol Chem* 272: 22053–22058.

18. Iwai K, *et al.* (2007) Expression and function of transmembrane-4 superfamily (tetraspanin) proteins in osteoclasts: reciprocal roles of tspan-5 and NET-6 during osteoclastogenesis. *Allergology Int* 56: 457–463.

19. Kacena MA, Gundberg CM, Horowitz MC (2006) A reciprocal regulatory interaction between megakaryocytes, bone cells, and hematopoietic stem cells. *Bone* 39: 978–984.

20. Kashida K, *et al.* (2000) Aquaporin adipose, a putative glycerol channel in adipocytes. *J Biol Chem* 275: 20896–20902.

21. Lacasa D, Garcia Dos Santos E, Guidicelli Y (2001) Site-specific control of rat preadipocyte adipose conversion by ovarian status: possible involvement of CCAAT/enhancer-binding protein transcription factors. *Endocrine* 15: 103–110.

22. Li J, Kuter DJ (2001) The end is just the beginning: megakaryocyte apoptosis and platelet release. *Int J Hematol* 74: 365–374.

23. Lin Y, Gill ME, Koubova J, Page DC (2008) Germ cell–intrinsic and –extrinsic factors govern meiotic initiation in mouse embryos. *Science* 322: 1685–1688.

24. Llamazares S, *et al.* (1991) Polo encodes a protein kinase homolog required for mitosis in Drosophila. *Genes Devel* 5: 2153–2165.

25. Lordier L, *et al.* (2008) Megakaryocyte endomitosis is a failure of late cytokinesis related to defects in the contractile ring and Rho/Rock signaling. *Blood* 112: 3164–3174.
26. Luo L, *et al.* (2007) Regulation of geminin functions by cell cycle-dependent nuclear-cytoplasmic shuttling. *Molec Cell Biol* 27: 4737–4744.
27. Ma D, *et al.* (2009) S6K is involved in polyploidization through its phosphorylation at Thr421/Ser424. *J Cell Physiol* 219: 31–44.
28. Nicklin P, *et al.* (2009) Bidirectional transport of amino acids regulates mTOR and autophagy. *Cell* 136: 521–534.
29. Nishitani H, Lygerou Z (2002) Control of DNA replication licensing in a cell cycle. *Genes to Cells* 7: 523–534.
30. Obunike JC, *et al.* (2001) Transcytosis of lipoprotein lipase across cultured endothelial cells requires both heparan sulfate proteoglycans and the very low density lipoprotein receptor. *J Biol Chem* 276: 8934–8941.
31. Onn I, *et al.* (2008) Sister chromatid cohesion: a simple concept with a complex reality. *Annu Rev Cell Dev Biol* 24: 105–129.
32. Patel SR, Hartwig JH, Italiano Jr. JE (2005) The biogenesis of platelets from megakaryocyte proplatelets. *J Clin Invest* 115: 3348–3354.
33. Pohl J, *et al.* (2005) FAT/CD36-mediated long-chain fatty acid uptake in adipocytes requires plasma membrane rafts. *Mol Biol Cell* 16: 24–31.
34. Rajkovic A, *et al.* (2004) NOBOX deficiency disrupts early folliculogenesis and oocyte-specific gene expression. *Science* 305: 1157–1159.
35. Raslova H, *et al.* (2007) Interrelation between polyploidization and megakaryocyte differentiation: a gene profiling approach. *Blood* 109: 3225–3234.
36. Ravid K, Lu J, Zimmet JM, Jones MR (2002) Roads to polyploidy: the megakaryocyte example. *J Cell Physiol* 190: 7–20.
37. Richard FJ (2007) Regulation of meiotic maturation. *J Anim Sci* 85: E4–E6.
38. Seong Y-S, *et al.* (2002) A spindle checkpoint arrest and a cytokinesis failure by the dominant-negative polo-box domain of Plk1 in U-2 OS cells. *J Biol Chem* 277: 32282–32293.
39. Suzuki A, Saga Y (2008) Nanos2 suppresses meiosis and promotes male germ cell differentiation. *Genes Devel* 22: 430–435.

40. Taghon T, *et al.* (2002) *HOX-A10* regulates hematopoietic lineage commitment: evidence for a monocyte-specific transcription factor. *Blood* 99: 1197–1204.

41. Tang C-H, *et al.* (2008) Bone-derived SDF-1 stimulates IL-6 release via CXCR4, ERK and NK-κB pathways and promotes osteoclastogenesis in human oral cancer cells. *Carcinogenesis* 29: 1483–1492.

42. Vigneron S, *et al.* (2004) Kinetochore localization of spindle checkpoint proteins: who controls whom? *Mol Biol Cell* 15: 4584–4596.

43. Wang J, Maldonada MA (2006) The ubiquitin-proteasome system and its role in inflammatory and autoimmune diseases. *Cell Molec Immunol* 3: 255–259.

44. Welburn JPI, Cheeseman IM (2008) Toward a molecular structure of the eukaryotic kinetochore. *Devel Cell* 15: 645–655.

45. Whitfield JF (2008) The solitary (primary) cilium — a mechanosensory toggle switch in bone and cartilage cells. *Cell Signalling* 20: 1019–1024.

46. Wolber E-M, Jelkmann W (2002) Thrombopoietin: the novel hepatic hormone. *News Physiol Sci* 17: 6–10.

47. Wright LM, *et al.* (2005) Stromal cell-derived factor-1 binding to its chemokine receptor CXCR4 on precursor cells promotes the chemotactic recruitment, development and survival of human osteoclasts. *Bone* 36: 840–853.

48. Yagi M, Roth GJ (2006) Megakaryocyte polyploidization is associated with decreased expression of polo-like kinase. *J Thrombosis & Haemostasis* 4: 2028–2034.

49. Yagi M, *et al.* (2006) DC-STAMP is essential for cell-cell fusion in osteoclasts and foreign body giant cells. *J Exp Med* 202: 345–351.

50. Young, J. K. (2002) Anatomical relationship between specialized astrocytes and leptin-sensitive neurons. *J Anatomy* 201: 85–90.

Chapter 5

HOW DO LYMPHOCYTES AND OTHER BLOOD CELLS PROTECT THE BODY FROM HARM?

The skin is the main organ that protects the body from dangerous items in the environment. However, this protective barrier can be breached, when we are stung by wasps, bitten by dogs, or scraped by objects that tear the skin. These events can introduce harmful venoms or chemicals into the body, or can provide an entrance for the teeming millions of microorganisms that lurk upon the surface of the skin and which regard the body as a tempting mass of nutrients suitable for supporting their lives and reproduction. A defense against these challenges is provided by blood cells that migrate from the bloodstream to the site of injury.

Most (99%) of the cells of the blood are red blood cells that are essentially containers of hemoglobin. These cells have a remarkably simple anatomy and lack nuclei and other organelles. The remaining nucleated blood cells are the ones that have a defensive function. They can be considered one by one (Fig. 5.1).

1. B-Lymphocytes

B-Lymphocytes were originally named for an organ in chickens, the Bursa of Fabricius, in which they were initially studied. In humans, it is equally convenient to remember that they originate in and largely mature in the Bone Marrow. Lymphocytes in general have a rather uninteresting anatomy: they have a small, round shape with a sparse cytoplasm and a condensed, round nucleus. In spite of their unimpressive appearance, they have an ability that has puzzled and

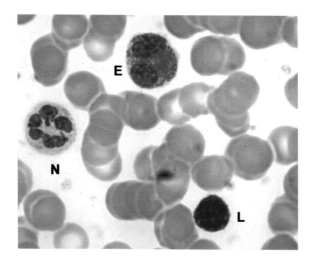

Fig. 5.1. A blood smear, showing red blood cells and nucleated cells that protect the body. These include a lymphocyte (L), neutrophil (N) and an eosinophil (E).

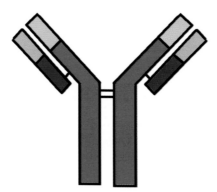

Fig. 5.2. Diagram of an *immunoglobulin* molecule. Relatively constant regions of this molecule are shown in red and blue; highly variable regions of this molecule are shown in pink and light purple.

intrigued scientists for fifty years. This ability is to produce a protein — an *immunoglobulin* or antibody — that shows an incredible variability between one lymphocyte and another.

An *immunoglobulin* (*Ig*) is a Y-shaped molecule composed of four protein subunits (Fig. 5.2).

The stem of each Y-shaped immunoglobulin molecule contains strands of amino acids that show relatively little variability. As a matter of fact, each stem (or "constant" region) comes in only five slightly different types, or classes: so-called *IgA*, *IgD*, *IgE*, *IgG*, and *IgM*. These slightly varying forms of the *Ig* molecule permit the constant region to either be inserted into a cell membrane or to be secreted as part of a free molecule. These are relatively trivial differences between molecules.

The more variable regions of the *Ig* molecule contain sequences of about 110 amino acids that show an amazing capability for variation. These variable regions bind to foreign molecules, termed antigens. It has been estimated that lymphocytes of the body can make as many as 100 million to one billion different varieties of *Ig*![9] This variability allows each *Ig* molecule to potentially bind to almost any conceivable molecular shape, permitting lymphocytes to attach to any possible molecule that has invaded the body. It is this amazing variability in *Ig* molecules that is the foundation for the defense of the body.

However, this variability raises a number of disturbing questions. There are only about 25,000 actively transcribed genes in the human genome. How, then, can the DNA of a lymphocyte code for millions of different *Ig* molecules? Also, if *Ig*'s can bind to almost any conceivable molecule, why do lymphocytes not attack the molecules of our own body as well as invading molecules? These fundamental questions have occupied immunologists for the last fifty years. Answers to these questions have resulted from studies of how lymphocytes develop.

B-Cell development

B-cells originate in the bone marrow from hematopoietic stem cells. A subset of these cells matures into a cell type called a common lymphoid progenitor, which can differentiate into a number of cell types of the immune system. These cells are influenced by transcription factors that bind to DNA and regulate gene transcription. One of these has the rather unsavory name of *PU.1*, which stands for the *Purine*

Rich Box on DNA that it binds to. High levels of *PU.1* in a cell force the cell to differentiate into a *macrophage*; lower levels turn on the genes for lymphocytes. The level of *PU.1* in a cell, in turn, is regulated by another transcription factor, *IRF8* (this name stands for *Interferon Regulating Factor 8*). So, right from the beginning, a complex network of DNA-binding proteins guides the development of B-cells.[38]

As B-cells develop, they produce another transcription regulating protein called *Pax5*. *Pax5* suppresses the transcription of 110 genes that would shift the lymphocyte away from the proper path of development. One of these genes, *BLIMP* (B-lymphocyte induced maturation protein), was discussed in Chapter 1: *BLIMP* forces the lymphocyte to become an antibody-secreting plasma cell. This cannot be permitted in the early life of a lymphocyte, and so is blocked by *Pax5*.[5]

Immunoglobulin production

After these initial stages, a maturing B-cell is ready to begin production of its own *Ig* molecule. To do this, 2 large protein subunits (heavy chains) and 2 smaller protein subunits (light chains) must be produced according to the instructions provided by their corresponding genes. For each heavy chain, a stretch of amino acids representing a constant region must be added to other sequences that are more variable. These variable amino acid sequences are termed the Variable (V), Diversity (D), and Joiner (J) regions. Thus, each final heavy chain is composed of C, V, D, and J sequences of amino acids (Fig. 5.3).

The curious thing about the genes for these sequences is that chromosome 6 of each lymphocyte contains many different varieties of DNA sequences for each potential V, D, J or C region of the protein. There are about 300–1000 sequences coding for the V region, 13 sequences coding for the D region, 4 sequences coding for the J region, and as we have said, 5 possible sequences coding for the C region.

One possible way a lymphocyte could choose among all these sequences would be to have *RNA polymerase II* settle down at

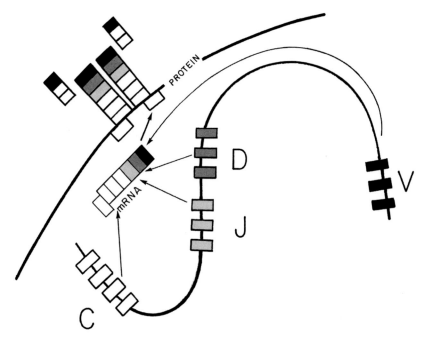

Fig. 5.3. Diagram showing how the heavy chains of *Ig* molecules are assembled from selected gene products for V, D, J, and C regions. Modified from Ref. 29.

random on one and only one of the 1000 V sequences and read that sequence, while ignoring all others. This logical mechanism, however, is never employed. Lymphocytes do something far more drastic to their DNA. They employ an enzyme called *RAG* (for *Recombination Activating Gene*) that recognizes stretches of 7 bases adjoining the V, J, or D sequences. When the enzyme randomly attaches to one of these adjoining sites, it forces the formation of a large loop of DNA and then cuts this loop completely out of the chromosome! This discarded loop will later be destroyed.[31]

Lymphocytes are the only cells that physically cut and splice their DNA in this radical way. If the *RAG* enzymes attacked other parts of the chromosomes, each lymphocyte would be destroyed; fortunately, they are activated only when bound to the specific flanking sequences within the complex of *Ig* genes. As a consequence of the effects of *RAG* enzymes, only a few V, J, or D sequences remain within the chromosome. Once these remaining sequences are translated into proteins, the assembly of the final heavy chains of the *Ig* molecule proceeds. A similar process yields the light chains of the finished *Ig* protein.

Once this bizarre process has been completed, each B-lymphocyte will display its own form of *Ig* molecule on the surface of the cell. This is accomplished by ensuring that the *Ig* molecule is a class M *Ig*, or *IgM*, which can insert the constant region of the Y-shaped molecule (called the *Fc region*) into the plasma membrane. Following this, a possibly dangerous event can occur: the *IgM* protein may attach to one of the normal proteins of the body that happens to be present in the bone marrow.

This event is potentially very hazardous, because it could lead to an auto-immune attack upon the body itself. In some uncertain way, autoreactive B-cells are instructed to avoid this danger. They are instructed to either a) undergo programmed cell death (a process called apoptosis, which will be reviewed later in this book) or b) activate the *RAG* enzymes all over again to produce a different type of *IgM*, a process called receptor editing.[13,21,27,36] This scrutiny of B-cells in the bone marrow diminishes the danger of the immune system attacking the body. The production of B-cells is a highly active process: in a mouse, 10 million B-cells are released from the bone marrow every day.[31]

After their release from the bone marrow into the bloodstream, B-lymphocytes migrate to lymph nodes and the spleen. Chemotactic factors produced by stromal cells of lymph nodes, including *stromal cell derived factor* and a molecule called *CXCL12*, help guide the cells to their targets.[3] B-cells, however, are by no means finished with their development upon reaching these targets.

When a B-cell arrives at a lymph node, membrane-bound *IgM* molecules may encounter the specific antigen that they are designed to bind to. If the antigen is small enough, the B-cell will internalize the antigen and digest it, since B-cells have a modest ability for phagocytosis. Fragments of the antigen will then be returned to the cell surface and displayed, or "presented," to other cells. If no other cells respond to this display of antigen, the B-cell is effectively thwarted: without the aid of other cells, it cannot proceed further in the defense of the body. What cells, then, help the B-cells?

B-lymphocyte activation

A completely different type of lymphocyte now enters the picture. These lymphocytes mature in the thymus, and are called helper T-cells. Helper T-cells also have highly variable receptors on their cell membranes. If a helper T-cell collides with an antigen-bearing B-cell in a lymph node, and if it has the right receptor for that specific antigen, it exerts powerful stimulatory effects upon the B-cell. How does it do this?

Lymphocytes have numerous membrane proteins that have been gradually identified over the years. These proteins are called *CD proteins* (*Cluster of Differentiation proteins*). As each protein was studied, it was assigned a number. Two such proteins are particularly important for B-cells: *CD40* and *CD80*. These B-cell proteins can be activated by complementary proteins on the surface of T-cells: *CD40* ligand and *CD28*. Thus, when a helper T-cell bearing these proteins, plus the receptor for the correct antigen, attaches to a B-cell, the B-cell explodes into activity.[23] This B-cell response is also aided by a protein, called *interleukin-4*, which is secreted by T-cells.

One particularly strange consequence of B-cell activation is that the B-cell begins to synthesize an enzyme called *activation-induced deaminase* (*AID*). The function of this enzyme is to induce point mutations in the DNA of the B-cell! It does this by removing the amino group from deoxycytidine in DNA, which changes deoxycytidine to deoxyuridine. Fundamentally, this changes the code on the DNA! Other DNA repair enzymes are attracted to this anomaly. One, called *UNG uracil-DNA glycosylase*, removes the aberrant uracil from the DNA, which creates a gap in the sequence (Fig. 5.4). Other enzymes repair this gap by replacing the missing nucleotides, often changing the sequence from a C to an A.[8,9] Thus, mutations are deliberately introduced into the sequences coding for *Ig* proteins.

The value of this strange procedure is that it introduces yet another level of variability into the structure of *Ig* proteins.

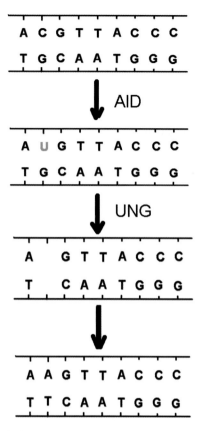

Fig. 5.4. Diagram of the process of DNA modification that takes place during somatic hypermutation in lymphocytes.

The genetic rearrangements initially produced by the *RAG* enzymes have been calculated to suffice for 100,000 to 1,000,000 different forms of *Ig* proteins; the additional variability introduced by somatic hypermutation raises the number of potential antibodies to between 100 million and a billion.[9] This is sufficient to provide antibodies that will bind almost any foreign protein. It is fortunate for the B-cell that this hypermutation of DNA only takes place at the genes for *Ig* proteins; if the rest of the cell's DNA were also to become mutated, chaos would ensue. At the same time that hypermutation takes place, the gene coding for the *IgM* molecule is converted to one containing a different "constant" region of the molecule, changing the *IgM* to an *IgG*.[15] This finally prepares the B-lymphocyte for its final fate, since *IgG* molecules are not membrane-bound and can be secreted.

Activation of a B-cell stimulates it to divide into hundreds of identical daughter cells called plasmablasts. These accumulate around the parent B-cell in a process called clonal expansion. The mass of these light-staining plasmablasts becomes visible as a pale structure within a lymph node called a germinal center (Fig. 5.5). Frequently, mitotic figures of the rapidly dividing plasmablasts can be seen within these centers. As the plasma cells mature, they leave the lymph nodes and secrete massive amounts of antibodies into their environment. These antibodies set the stage for an attack upon a foreign antigen (described later in this chapter). Some of the activated B-cells fail to turn into plasma cells, but instead remain as long-lived memory cells. These will circulate throughout the body and can respond to a specific antigen many years after the initial insult.

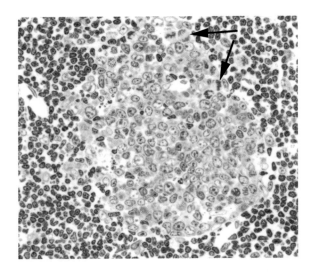

Fig. 5.5. A germinal center in a lymph node, showing the light-staining plasmablasts surrounded by a halo of dark-staining, mature lymphocytes. Several mitotic figures can be seen (arrows).

2. T-Lymphocytes

This subset of lymphocytes also originates in the bone marrow, but then migrates to the thymus where critical stages of maturation occur. Our understanding of the thymus has undergone dramatic revisions over the last forty years.

As late as 1963, the thymus was regarded as an uninteresting and unimportant organ. It is large in newborn animals, but its growth does not keep pace with the rest of the body, and by adulthood much of its substance becomes replaced by fat. When lymphocytes from an adult thymus were examined *in vitro*, they were found to be incapable of producing any antibodies. Also, when the thymus was removed from adult animals, no immunological deficits were detectable. These findings led many scientists to conclude that the thymus was basically useless as an immune organ.

These conclusions were challenged by two events that occurred during the 1960's: 1) other investigators removed the thymus from newborn animals and found that such newborns had a very impaired ability to produce antibodies, in spite of having normal amounts of B-cells. 2) another finding came from a lab in Scotland, which reported discovering a peculiar mutation in mice. These mice had abnormal hairs that failed to grow beyond hair follicles in the skin, giving the impression that the mice were hairless, or "nude." Nude mice resemble little mouse-shaped pink rubber erasers. Also, these mice failed to develop a normal thymus and were severely immunocompromised. In fact, tissues from other breeds of mice could be grafted onto nude mice and failed to elicit a normal immune response that should have led to rejection and destruction of the foreign tissue.[24] Since then, the thymus has come to be viewed as far from boring, and in fact contains some of the most peculiar and interesting cells in the body. How does the thymus promote the differentiation of lymphocytes? What is wrong with the nude mice?

Reticulo-epithelial cells of the thymus

It is now clear that the deficit in nude mice results from a mutation in a DNA-binding transcription factor called *Foxn1*, which is normally expressed only in the epithelial cells of the skin and in the endoderm-derived cells of the thymus.[24] This explains the deficient thymic function of these animals, because the normal structure of the thymus, unlike that of most organs, is highly dependent upon epithelial cells. These cells, termed reticulo-epithelial cells, behave like no other epithelial cell in the body. They do not line hollow structures like ducts or blood vessels, but instead are distributed among the masses of lymphocytes found in the thymus (Fig. 5.6).

Reticulo-epithelial cells contain keratin intermediate filaments like other epithelial cells and also have junctional complexes that bind the cells to each other, but these junctions are located at the tips of long cell processes stretching away from the cell area containing the nucleus. These cells make up a supporting framework for the thymus; in other organs, most of the internal structure is provided by connective tissue

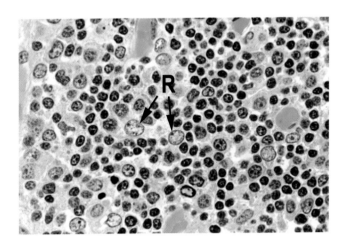

Fig. 5.6. Thymic reticulo-epithelial cells (R) have pale nuclei and can be identified among surrounding masses of lymphocytes.

fibroblasts and collagen, rather than by a peculiar arrangement of epithelial cells. Reticulo-epithelial cells also have a critical role in regulating the development of T-cells.

Every day, about 100 pre-T cells reach the thymus from the blood stream. They enter thymic blood vessels that are located at the junction between the outer cortex of the thymus and the innermost medulla. Once inside the thymus, each cell divides about 20 times, producing about 50 million daughter cells daily. These daughter cells begin migrating up towards the outer cortex of the thymus. However, only about 5% of them (2 million cells) will ever complete their journeys to re-enter the medulla and leave the thymus.[20] Most of the pre-T cells of the thymus are destroyed! This process is necessary, because it eliminates many of the T-cells that would attack the body.

Mature T-cells have receptors on their cell membranes, called T-cell receptors, which closely resemble immunoglobulin molecules. Like immunoglobulins, these receptors have highly variable regions, and are produced in T-cells in a process much like that seen in B-cells that produce immunoglobulins. Like in the bone marrow, major tasks for the thymus are 1) the identification of T-cells with receptors that could attack foreign antigens and 2) the elimination of T-cells with receptors that could attack the native proteins of the body.

When pre-T cells enter the thymus, they encounter reticulo-epithelial cells that govern their development. Reticulo-epithelial cells produce growth factors like interleukin-7 (IL-7), which greatly stimulates T-cell proliferation. IL-7 also stimulates the production of recombinase enzymes that scramble the DNA of pre-T cells and leads to the production of highly variable T-cell receptor proteins, much like the process for *Ig* production in B-cells. In addition, reticulo-epithelial cells possess a membrane-bound protein called Stem Cell Factor that is critical for T-cell development; if this is deleted experimentally, mice undergo a 10-fold reduction in the development of T-cells.[32]

Perhaps the most important proteins expressed by reticulo-epithelial cells are called *MHC proteins*, named for the *Major Histocompatability Complex* of genes that code for them. There are two classes of *MHC* proteins: *MHC I* proteins, that are particularly abundant on lymphocytes but are found in trace amounts on most nucleated cells of the body, and *MHC II* proteins, that are mainly found on lymphocytes, antigen-presenting cells like macrophages, and thymic epithelial cells.

An *MHC I* protein is composed of 3α subunits, 1β subunit and a so-called C region that is anchored into the cell membrane. *MHC I* proteins can vary greatly between one individual and another. This is because the genes for the 5 subunits are different from person to person. About 59 different "alleles," or variations in the α subunit, have been found in the human population, while 111 different varieties of the β subunit exist.[31] These subunit types can be combined in many thousands of ways, making it possible to define thousands of different types of human beings on the basis of their *MHC I* proteins. When organ transplants are performed, it is important to perform "tissue matching" procedures to determine that the *MHC I* protein of a donor is as similar as possible to the *MHC I* protein of an organ recipient.

In addition to signaling the immune system that a cell "belongs" to the body, the *MHC I* protein also functions in the process of antigen presentation. If a virus has infected a cell, some viruses will be degraded by proteasomes into small peptide fragments. These peptide fragments (antigens) will be moved to the cell membrane and attached to an *MHC I* protein. If a T-cell can bind to both the *MHC I* protein and the antigen, it will attack the infected cell. A major job of the thymus is to produce T-cells, called cytotoxic T-cells, which can perform this function.

MHC class II proteins are composed of two different protein subunits and are particularly important for the production of antibodies, as we shall see. But what role do *MHC* proteins play in T-cell development?

Thymic nurse cells

It appears that developing T-cells with receptors having a high avidity for the *MHC* proteins are potentially dangerous attackers of the body; these somehow are stimulated to undergo programmed cell death (apoptosis).[32] A good proportion of this selection process for T-cells

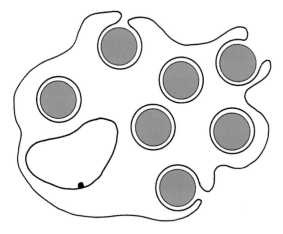

Fig. 5.7.   Diagram of a thymic nurse cell. Lymphocytes engulfed within the thymic nurse cell are diagrammed in green.

appears to take place *within the cytoplasm* of very peculiar reticulo-epithelial cells called thymic nurse cells (Fig. 5.7).

These cells can engulf between 20 to 200 maturing lymphocytes within a single cytoplasm! A subset of the engulfed cells proliferates and matures further to finally exit the nurse cell; other types of maturing lymphocytes are induced to undergo programmed cell death. The importance of these peculiar cells for thymic function can be demonstrated by growing a thymus in culture and then specifically damaging the nurse cells with an antibody to them: this results in an 80% decline in the release of T-cells from the thymus.[14] The reasons why this strange intracellular environment is necessary for T-cell development are still not known.

Medullary reticulo-epithelial cells

Following an initial selection process in the upper cortex of the thymus, T-cells move towards the medulla of the thymus. There, they encounter yet another very strange type of reticulo-epithelial cell. These medullary cells express a protein called *Aire* (an abbreviation of *Autoimmune Regulator protein*). This protein has the capacity to bind to histone H3 in the DNA of these cells.[18] In some manner, this permits *Aire* to stimulate the transcription of about 2000 proteins that otherwise would never be expressed in the thymus. This huge array of proteins, amounting to about 10% of all the proteins coded for by the human genome, represents a sampling of proteins that are normally components of only specific tissues.[6] Proteins normally found only in the liver, muscles, pancreas, ovaries, lungs, brain, etc., have now been found to be expressed in medullary thymic reticulo-epithelial cells, under the control of the *Aire* protein.

When a developing T-cell reacts to any of these proteins, it is also suppressed by cell death or via other mechanisms. Thus, the thymus, by presenting a "preview" of bodily proteins to T-cells, prevents most auto-reactive T-cells from ever exciting the thymus and attacking the body. Recently, it has been found that *Aire*-expressing cells also exist in lymph nodes and finish the job by eliminating autoreactive T-cells that had somehow escaped the thymus unscathed.[12]

Mature T-lymphocytes

Even after T-cells have been released from the thymus, they are by no means ready to achieve their mature functions. Like B-cells released from bone marrow, T-cells require help from other cell types. Two main types of T-cells exist: helper T-cells and cytotoxic T-cells.

Helper T-cells become activated when they interact with cells of the macrophage lineage. Macrophages and their relatives, dendritic cells that populate lymph nodes, are professional phagocytes that are descended from blood cells called monocytes. An understanding of their phagocytic ability is complicated by the question of how do macrophages "know" what to phagocytize? How do they distinguish between normal molecules in their environment and harmful, foreign molecules?

The answer is that macrophages have a number of cell-surface receptor proteins that bind foreign molecules and stimulate the macrophage to ingest them and process them in lysosomes. One such receptor is called a scavenger receptor, which binds particularly well to components of bacterial cell walls.[35] After a bacterium has been phagocytized and partially digested in lysosomes, peptide fragments of the bacterium are returned to the cell membrane and are presented to T-cells in the environment.

Within a lymph node, T-cells are highly mobile and pass rapidly across the surface of a dendritic cell. It has been estimated that a single dendritic cell may come in brief contact with as many as 5000 T-cells per hour![25] This allows for an efficient identification of just the right T-cell for a given antigen. The process of antigen presentation by macrophages to T-cells can be visualized by injecting India ink into the footpad of a rat. The ink will be carried by lymphatic vessels to the popliteal lymph nodes behind the knee, where it will be phagocytized by macrophages (Fig. 5.8).

If a helper T-cell possesses a receptor that will bind the antigen displayed by the macrophage, the two cell types gradually form a close association. Over 12 hours, a T-cell will make multiple, hour-long contacts with the macrophage, forming structures called "immunological synapses." What molecules comprise these zones of cell-cell contact?

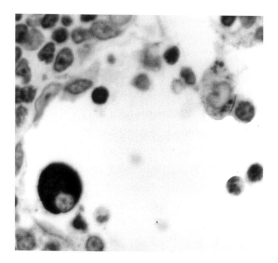

Fig. 5.8. View of the subcapsular sinus of a lymph node, showing a macrophage with a pale, bean-shaped nucleus and brown cytoplasmic inclusions formed by India ink particles. Some of the particles are displayed on the cell surface and are being presented to a lymphocyte. A dark-staining mast cell is visible in the lower left-hand corner of this picture.

Immunological synapses

When a macrophage presents an antigen such as a fragment of a bacterium to a helper T-cell, the antigen is held within a cleft formed by a *MHC class II* protein on its cell surface. The T-cell binds to these proteins via the variable portion of its T-cell receptor, plus a co-receptor protein called *CD4* (Fig. 5.9). This binding event activates the cell and sets in motion a number of very interesting processes.

The full activation of a helper T-cell by a macrophage requires about 24 hours. During this time, the T-cell develops an asymmetry in cell proteins, with some proteins remaining close to the "immunological synapse," and other proteins moving towards the other half of the cell. Finally, the T-cell is stimulated to divide. The half of the cell

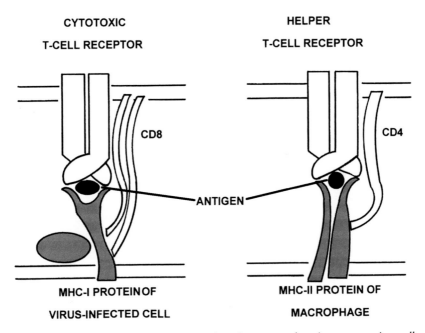

Fig. 5.9. Diagram of binding between lymphocytes and antigen-presenting cells. On the right, a helper T-cell binds to an antigen-presenting macrophage via its T-cell receptor and the CD4 co-receptor protein. On the left, a cytotoxic T-cell binds to a virus-infected cell via its T-cell receptor and the CD8 co-receptor protein. Modified from Ref. 16.

in contact with the macrophage forms a new daughter cell. This daughter cell becomes a "memory" T-cell, which is not further activated, but which can persist for decades and "remember" the antigen that was presented.

The existence of "memory" T cells, incidentally, has proven to be the explanation for all of those puzzling experiments involving the thymus in the 1950's. These experiments showed that removing the thymus in adult animals had no effect on immune function. The reason for this was that "memory" T cells, produced early in life, could persist and provide a substitute for an intact thymus, generating helper T cells even when the thymus was no longer present.

The second half of the initial cell farthest from the "immunological synapse" turns into a second type of daughter cell, a fully activated helper T-cell that will soon become completely active. This asymmetrical type of cell division, with 2 different cell fates for 2 distinct daughter cells, solves a problem encountered by most tissues. This problem is how to produce more cells without disrupting the highly differentiated function of a tissue. As we noted at the beginning of Chapter 3, the solution to this problem is to perform an asymmetrical type of cell division, with a long-lived stem cell forming one daughter cell and a shorter-lived, fully differentiated cell forming the other daughter cell. This is essentially what happens when helper T-cells become activated by dendritic cells or macrophages.[4]

Within 5 days of the meeting between a T-cell and a macrophage, a single T-cell may divide 10 times and produce over one thousand daughter cells. These then fly to the aid of B-cells and stimulate them to become antibody-producing plasma cells.

In spite of all this discussion about antibodies, we still have not clarified their utility. What are antibodies good for? The answer is simple: antibodies can bind to, or "opsonize" almost any foreign antigen in a very specific way. This leads to the destruction of the antigen, because macrophages have receptors for the constant region of antibody molecules. Hence, once an antigen is coated with antibodies, it will be phagocytized by macrophages and destroyed.

Cytotoxic T-cells

Cytotoxic T-cells also respond to antigen presentation by cells, but the circumstances and the cellular responses differ from those of helper T-cells. Cytotoxic T-cells bind to *MHC I* proteins on a cell surface, aided by the *CD8* co-receptor, rather than binding to *MHC II* proteins. Also, they frequently react to the presence of viral proteins on the surface of an infected cell. Cytotoxic T-cells respond by killing the infected cell to protect the surrounding, uninfected cells. Cytotoxic T-cells have a number of weapons in their arsenal that they use to induce cell death.

One T-cell weapon is a membrane protein called *FAS ligand*. It binds to a target cell membrane protein called *FAS*. This membrane protein, when bound by *FAS ligand*, activates a cytoplasmic protein sequence called a "death domain." This domain initiates the process of apoptosis.

Other T-cell weapons consist of two secreted proteins, *perforin* and *granzyme B*. Current data suggest that both proteins are released from T-cells via *exocytosis*, and then are taken up into a target cell via *endocytosis*. Once inside a target cell, *perforin* becomes inserted into endosomal membranes and causes permeable pores to form. This, in turn, releases *granzyme B* from the endosomes into the cytoplasm. *Granzyme B* then initiates apoptosis via a number of complementary pathways.[28]

Infection of helper T-cells by the HIV virus

A viral infection that is particularly devastating to the immune system is carried out by the Human Immunodeficiency Virus (HIV, or AIDS virus). This virus possesses two surface glycoproteins, called *gp120* and *gp 41*, which interact with the *CD4* protein and a cytokine receptor called *CCR5* found on the surface of helper T-cells, macrophages, dendritic cells, and microglia.[39] This interaction allows the virus to specifically target and infect these immune cells. In addition to directly damaging helper T-cells, the viral infection attracts the unwelcome attention of cytotoxic T-cells, which proliferate and attempt to destroy infected helper T-cells. All of these effects severely compromise the ability of the immune system to resist infections by other organisms. Approximately 25 million people have been killed worldwide by this disease since its outbreak from selected regions in Africa in the 1970's.

RNA interference and defense against viruses

The immune system may also employ a weapon more ancient than cytotoxic T-cells to destroy viruses. This weapon is called *RNA interference*, and is seen in almost all living things, from plants to yeasts to insects. Viruses are simple protein capsules that contain genetic material in a number of forms: single-stranded DNA or single- or double-stranded RNA. A virus injects this genetic material into a cell and either viral or cellular polymerases make multiple copies of the nucleic acids, which are then packaged into hundreds of new viruses that kill the cell when they are released. Apart from killing the infected cell, how can the body defend itself against viruses?

RNA interference is a non-lethal mechanism that requires a number of different proteins. First, when double-stranded RNA from a virus appears within a cell, it is detected by an enzyme called *Dicer* and cleaved into small (21 bases long) fragments called *small interfering RNAs (siRNAs)*. Each one of these fragments becomes an active part of a multi-protein complex called *RISC (RNA-induced silencing complex)*. The *siRNA* in each *RISC* complex hybridizes with the viral RNA in each cell and guides the destruction of its complementary viral RNA. This is an effective way of preventing RNA from having effects in a cell and may have preceded more complex immunological mechanisms during evolution.[37]

Recently, attempts have been made to utilize *siRNA* to combat the HIV virus. It turns out that all T-cells possess a membrane protein called *CD7*. The function of this protein is uncertain, and it may not be required for T-cell health. However, it offers a way to deliver molecules specifically to T-cells. In a new approach, investigators administered an antibody to the *CD7* protein to mice. Attached to this antibody was a strand of antiviral *siRNA*. When the antibody bound to the *CD7* protein, it was internalized into T-cells, dragging the *siRNA* along with it. This allowed the *siRNA* to suppress the replication of the HIV virus in T-cells.[19] This seems to be a very promising way to defeat the devastating effects of HIV infection.

Other types of lymphocytes (natural killer cells, suppressor cells, etc.) can be identified in the body, but a complete description of all of them would require these passages to veer completely into the realm of immunology. Let us turn our attention, then, to other types of blood cells.

3. Neutrophils

Neutrophils are large white blood cells having several striking characteristics: each cell has a distorted nucleus showing a number of constrictions that produce 2–3 lobules for each nucleus, and all neutrophils have abundant, small vesicles in the cytoplasm that fail to stain intensely with either basic or acidic stains (hence the term "neutrophil," as a description of the cell's cytoplasm) (Fig. 5.1).

The cytoplasmic granules of neutrophils consist of small lysosomes (primary granules) or larger secretory (secondary) granules that contain a protein called *lactoferrin*. This protein, related to a blood-borne protein called *transferrin*, has a high affinity for iron; by sequestering iron, a needed element for cell growth, *lactoferrin* has a potent antibacterial activity. Neutrophils also kill bacteria by phagocytizing them and then fusing the phagocytic vesicle with lysosomes. The lysosomes deposit hydrolytic enzymes onto the bacteria and also attack them with hydrogen peroxide.[11,16]

The cytoplasmic vesicles of neutrophils, while specialized, are not so different from vesicles of many other cells. It is the strange shape of the nucleus that attracts the attention of cell biologists. How does the nucleus become so distorted, and what function might this serve? What causes these cells to acquire their strange appearance?

The differentiation of pre-neutrophil cells (called promyelocytes) is stimulated by capillary cells of bone marrow, which secrete a growth factor called *Granulocyte-Macrophage Colony Stimulating Factor* (*GM-CSF*). This stimulates the production of the cytoplasmic vesicles of neutrophils. Transformation of the nucleus, however, is stimulated by another signaling molecule — retinoic acid — that is sufficient to re-shape the nucleus from a bean-shaped structure into a lobulated one.[26]

This nuclear transformation results from two things. First, a drastic decrease in amounts of nuclear *lamin* proteins occurs in neutrophils. Since these types of intermediate filaments form links between nuclear pores and govern the overall shape and strength of the nuclear envelope, this change in *lamins* produces a more fragile and easily shaped nucleus. Second, cytoplasmic microtubules move towards the nuclear envelope and exert pressure on it that remodels the shape of the nucleus. Neither of these events are direct consequences of retinoic acid binding to the nucleus, but reflect changes in DNA transcription following exposure to retinoic acid.

Retinoic acid moves to the nucleus from the cytoplasm and binds to proteins called *retinoic acid receptors*. These receptors function as transcription factors that stimulate the transcription of other genes. Some of these genes we have met before. Retinoic acid induces the transcription of the *C/EBP* family of transcription factors. Some of these, as we have noted, are essential for the morphogenesis of fat cells (Ch. 4). During hematopoeisis, these proteins are also required for the genesis of neutrophils; if they are disabled in a mouse, the mouse will produce normal amounts of lymphocytes, monocytes, and red blood cells, but will completely lack any neutrophils or eosinophils. Additional transcription factors such as *PU.1* and *SP1* are also required for neutrophil differentiation.[10]

Why is it necessary to possess such an oddly-shaped nucleus? In rodents, neutrophil nuclei are even more peculiar and are actually annular (ring-shaped)! The basic assumption of most researchers has been that this odd shape makes it easier for neutrophils to force their way between endothelial cells of capillaries and enter the connective tissue to do battle with bacteria. This has not been easy to verify experimentally.

Another possible explanation for the strange nucleus of a neutrophil has been proposed only recently. Frequently, when neutrophils encounter bacteria, they produce so-called *Neutrophil Extracellular Traps* (*NETs*). These are formed when a neutrophil nucleus rapidly loses its shape and ruptures; following this, the cell membrane breaks, causing a mass of nuclear chromatin and cytoplasmic proteins to pour out of the cell. This mass of molecules is extremely sticky and harmful to bacteria.[11]

Many of us may remember these properties of DNA from lab classes in biochemistry. I recall an exercise that involved isolating DNA from tubes full of yeast. At the final step of isolation, the DNA coalesced in a test tube as a white mass of fine fibers very similar to a small piece of cotton candy and just as sticky. Neutrophils, then, can sacrifice the complex structure of nuclear DNA so they can use it as a simple mop to clean up bacteria.

4. Eosinophils

Eosinophils (Fig. 5.1) derive their name from their large cytoplasmic granules that stain intensely after exposure to the acidic dye, eosin. They account for only 2–4% of the white blood cells of the blood (neutrophils are the most numerous white blood cells and constitute 60% of all leukocytes). Like neutrophils, eosinophils have a lobulated nucleus (usually two lobes). The contents of the cytoplasmic granules of eosinophils are the weapons with which they fight disease.

The so-called *Major Basic Protein* of eosinophil granules is what accounts for the basic pH of the granules and the intense staining with the acid stain, eosin. This protein accumulates into distinctive crystalloid inclusions within each granule; its main function is to bind to and kill bacteria and the larvae of helminthic worms (roundworms) that have invaded the body.[1] This may not strike you, the reader, as a particularly pressing health problem, but in fact, in many parts of the world, about 50% of children running around barefoot contract soil-transmitted helminthic infections.[17] The *major basic protein* of eosinophils is a major weapon against these infections.

Other proteins secreted by eosinophils are also important. These include *eosinophil cationic protein, eosinophil peroxidase,* and an *eosinophil-derived neurotoxin.* All of these proteins have anti-worm and anti-bacterial properties. In addition, *major basic protein* may be harmful to normal human tissues and can cause the degranulation of mast cells.[1] Also, *VEGF* and *FGF* produced by eosinophils can increase blood vessel growth and enhance processes of inflammation. Eosinophils may exert harmful influences upon allergic diseases such as asthma.[29]

One final peculiar feature of eosinophils is that, when activated by the presence of chemicals from bacterial cell walls, they rapidly extrude ("catapult") the contents of their mitochondria into their environment. The DNA released from mitochondria traps and damages bacteria.[40] This may serve as a reminder of the origin of mitochondrial DNA: mitochondria are believed to have originated as bacteria that become symbionts within ancient, primitive cells. Thus, mitochondria retain a bacterial form of their own DNA. It seems ironic that the mitochondria of eosinophils, distant relatives of bacteria, could be utilized to kill external bacteria.

5. Mast Cells and Basophils

Both mast cells and basophils are noteworthy for the large, intensely stained membrane-bound vesicles (granules) that they carry within the cytoplasm (Fig. 5.10). Mast cells originate in bone marrow, travel briefly in the bloodstream, and then quickly take up residence within connective tissue. Basophils are rare blood cells, representing less than 1% of circulating leukocytes. They are similar to mast cells in that they have similar cytoplasmic granules, but differ because they have a lobulated rather than an ovoid nucleus. Upon stimulation, they too will leave the bloodstream and enter connective tissue.

Both cell types function when they are stimulated to degranulate and explosively release all the contents of their vesicles into the external environment of the cell. The degranulation of these cells takes place when the cells encounter an antigen (typically, for hay fever sufferers, the antigen may originate from tree pollen). Mast cells can react to an antigen because they possess membrane receptors for *Immunoglobulin E*. When the *IgE* molecules trapped on the surface of the mast cell bind to an antigen, this drags the membrane *IgE* receptors together into a mass, or so-called raft, on the membrane. This in turn attracts the attention of an enzyme called *IκB* kinase,

children from semi-urban communities in Nigeria: a double-blind placebo-controlled randomised trial. *BMC Infect Dis* 9: 20.

18. Koh AS, *et al.* (2008) Aire employs a histone-binding module to mediate immunological tolerance, linking chromatin regulation with organ-specific autoimmunity. *Proc Natl Acad Sci USA* 105: 15878–15883.

19. Kumar P, *et al.* (2008) T cell-specific siRNA delivery suppresses HIV-1 infection in humanized mice. *Cell* 134: 577–586.

20. Kyewski B, Klein L (2006) A central role for central tolerance. *Annu Rev Immunol* 24: 571–606.

21. Lang J, *et al.* (1996) B cells are exquisitely sensitive to central tolerance and receptor editing induced by ultralow affinity, membrane-bound antigen. *J Exp Med* 184: 1685–1697.

22. Lindstedt L, Lee M, Kovanen PT (2001) Chymase bound to heparin is resistant to its natural inhibitors and capable of proteolyzing high density lipoproteins in aortic intimal fluid. *Atherosclerosis* 155: 87–97.

23. Lumsden JM, Williams JA, Hodes RJ (2003) Differential requirements for expression of CD80/86 and CD40 on B Cells for T-dependent antibody responses *in vivo*. *J Immunol* 170: 781–787.

24. Mecklenberg L, Tychsen B, Paus R (2005) Learning from nudity: lessons from the nude phenotype. *Exp Dermatol* 14: 797–810.

25. Miller MJ, *et al.* (2004) T cell repertoire scanning is promoted by dynamic dendritic cell behavior and random T cell motility in the lymph node. *Proc Natl Acad Sci USA* 101: 998–1003.

26. Olins AL, Olins DE (2004) Cytoskeletal influences on nuclear shape in granulocytic HL-60 cells. *BMC Cell Biol* 5: 30–48.

27. Opferman JT (2008) Apoptosis in the development of the immune system. *Cell Death Differentiation* 15: 234–242.

28. Pipkin ME, Lieberman J (2007) Delivering the kiss of death: progress on understanding how perforin works. *Curr Opinion Immunol* 19: 301–308.

29. Puxeddu I, *et al.* (2005) Mast cells and eosinophils: A novel link between inflammation and angiogenesis in allergic diseases. *J Allergy Clin Immunol* 116: 531–536.

30. Qi H, *et al.* (2008) SAP-controlled T-B cell interactions underlie germinal centre formation. *Nature* 455: 764–769.

31. Rao CV (2006) Immunology. 2nd Ed., Alpha Science International: Oxford.
32. Savino M, Dardenne M (2000) Neuroendocrine control of thymus physiology. *Endocr Rev* 21: 412–443.
33. Scudamore CL, *et al.* (1998) The rat mucosal mast cell chymase, RMCP-II, alters epithelial cell monolayer permeability in association with altered distribution of the tight junction proteins ZO-1 and occludin. *Eur J Cell Biol* 75: 321–330.
34. Suzuki K, Verma IM (2008) Phosphorylation of SNAP-23 by IκB kinase 2 regulates mast cell degranulation. *Cell* 134: 485–495.
35. Thomas CA, *et al.* (2000) Protection from lethal gram-positive infection by macrophage scavenger receptor-dependent phagocytosis. *J Exp Med* 191: 147–155.
36. Tiegs SL, Russel DM, Nemazee D (1993) Receptor editing in self-reactive bone marrow B cells. *J Exp Med* 177: 1009–1020.
37. Van Rij RP, Andino R (2006) The silent treatment: RNAi as a defense against virus infection in mammals. *Trends Biotechnol* 24: 186–193.
38. Wang H, *et al.* (2008) IRF8 regulates B-cell lineage specification, commitment, and differentiation. *Blood* 112: 4028–4038.
39. Wyatt R, Sodroski J (1998) The HIV-1 envelope glycoproteins: fusogens, antigens, and immunogens. *Science* 280: 1884–1888.
40. Yousefi S, *et al.* (2008) Catapult-like release of mitochondrial DNA by eosinophils contributes to antibacterial defense. *Nature Med* 14: 949–953.

Chapter 6

GLIAL CELLS — THE UNSUNG HEROES OF THE BRAIN

Approximately 60% of the cells in the brain are neurons and the remaining 40% of cells are non-neuronal glia (astrocytes, oligodendroglia, and microglia) plus the endothelial cells of blood vessels.[13] While the existence of glial cells has been known for a long time, the roles of these cells in brain function have not, until recently, received as much attention as those of neurons. This chapter is a brief attempt to somewhat rectify this neglect of glia.

1. Astrocytes

Astrocytes account for about 30% of the total volume of most brain areas.[45] They can be easily identified by their pale-staining, oval nuclei and also by their possession of a special type of cytoplasmic intermediate filament composed of *glial fibrillary acidic protein* (*GFAP*) (Fig. 6.1).

Fine processes, supported by the tough filaments of *GFAP*, radiate out from each astrocyte. In rats, the mass of astrocytic processes extending from a single astrocyte may contact 600 neuronal dendrites and 100,000 synapses; in humans, this figure is even larger.[20] Astrocytes in the cerebral cortex of humans acquire more complex shapes than those of other animals, leading to speculation that specialized astrocytes of humans make a considerable contribution to the enhanced abilities of the human cerebral cortex.[29]

Astrocytes develop in restricted regions of the neural tube, under the influence of a helix-loop-helix type transcription factor called *SCL*.[26] Homeotic proteins such as *Pax6* and *NKx6.1* also influence

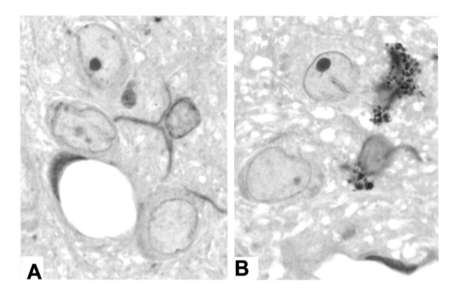

Fig. 6.1. Two examples of astrocytes. On the left, an astrocyte with a pale-staining nucleus and thin, GFAP-containing processes touching three adjacent neurons, which possess larger nuclei and prominent nucleoli. The darker-staining nucleus of a small oligodendrocyte is also visible. On the right, two Gomori-positive astrocytes with dark-staining cytoplasmic granules are shown.

their development.[15] New findings show that astrocytes have a wide range of activities that are critical for the function of the brain.

Astrocytes and brain blood vessels

The brain is a highly vascular structure, but the flow of blood and the behavior of blood vessels change considerably from region to region, depending upon the degree of activity of a given brain area. When neurons become highly activated, they require a greater supply of oxygen and nutrients to maintain their activation and blood vessels become dilated to meet this demand. This change in blood vessel function in fact is the basis for a widely used technique, called functional MRI (magnetic resonance imaging), which can reveal patterns of activation in the brain of a living person. If, for example, a person is asked to focus on patterns of blinking lights, nerve cells in the visual cortex will be activated, and blood vessels will deliver more oxygenated hemoglobin to this region. Since oxygenated hemoglobin reacts to a magnetic field differently than deoxygenated hemoglobin, activated brain areas will "light up" during a fMRI scan.[25]

Blood vessels in activated areas do not, however, react to activated neurons directly. Instead, blood vessels are instructed to dilate or constrict by astrocytes. Astrocytes rapidly react to the release of neurotransmitters from neurons by shifting levels of intracellular calcium and by releasing metabolites of arachidonic acid from so-called "end feet," processes of astrocytes that terminate upon blood vessels. These cause a dilation of blood vessels and permit local increases in neuronal activity during a specific task.[20,43]

Astrocytes do not only regulate the flow of blood in vessels, but they also have a fundamental influence upon blood vessel permeability. Most brain capillaries are unusually impermeable, because their endothelial cells possess many tight junctions that form water-tight seals between cells. These junctions constitute the so-called "blood-brain barrier" that does not permit large molecules to enter brain tissue from the blood. It is now known that contact with astrocytes stimulates brain capillaries to maintain the blood-brain barrier.[44] An astrocyte protein called *SSeCKS* (*src-suppressed C-kinase substrate*) appears to be involved in inducing the blood-brain barrier in capillaries.[22]

This barrier has a helpful, protective function, but also poses some problems, as well. The brain is highly dependent upon glucose as a fuel molecule, but glucose cannot passively diffuse into the brain across leaky capillaries, as it can in other tissues. To supply brain tissue with glucose, capillary endothelial cells are provided with a type of *glucose transporter protein* (called *GLUT1*) that has a high affinity for glucose. Glucose transported across capillary walls is then absorbed into cells via a variety of other types of glucose transporters (*GLUT1* for astrocytes and oligodendrocytes, *GLUT3* for neurons, and *GLUT5* for microglia).[39,50]

Certain brain regions are peculiar in that their capillaries are extremely leaky and *lack* a blood-brain barrier. These regions are termed circumventricular organs because of their proximity to

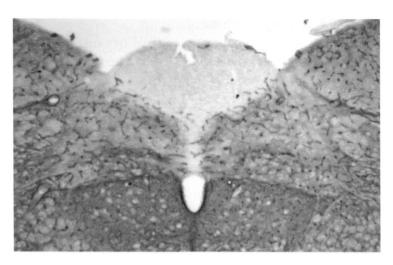

Fig. 6.2.   Capillaries of the triangle-shaped area postrema, unlike in the surrounding brain tissue of the dorsal medulla, are extremely leaky and fail to stain for *GLUT1* glucose transporter proteins.

the third and fourth ventricles, fluid-filled cavities within the brain.[33,52] An example of one of these regions is the area postrema, located on the dorsal surface of the medulla and below the fourth ventricle (Fig. 6.2). Capillaries in this triangular structure are extremely leaky and lack *GLUT1* glucose transporters.[46]

The area postrema functions as a "chemoreceptor" and monitors the presence of noxious or potentially dangerous molecules in the bloodstream. If such a molecule is detected, neurons of the area postrema signal to adjacent vagal and hypoglossal neurons and stimulate vomiting, which removes the offensive chemical from the body.

Another example of a circumventricular organ is the median eminence and the adjacent arcuate nucleus of the basal hypothalamus (Fig. 6.3). Neurons and astrocytes in this region are freely exposed to molecules circulating in the blood and possess receptors for these molecules.[8] The fact that astrocytes fail to induce blood-brain barrier properties in these brain regions suggests that circumventricular astrocytes are specialized, relative to other astrocytes. In addition, other features of circumventricular astrocytes are also of interest.

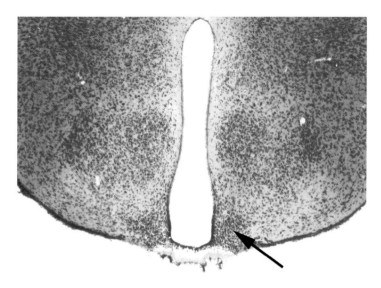

Fig. 6.3. View of the ventral hypothalamus, showing the arcuate nucleus (arrow) just above the median eminence and located on both sides of the third ventricle.

Sensory functions of astrocytes

Because circumventricular astrocytes are exposed to blood-borne molecules, they have the potential to function as monitors of the blood. And indeed, astrocytes near the median eminence and area postrema possess high capacity type glucose transporters (*GLUT2* transporters) that are identical to the ones used as glucose sensors by cells in the liver and Islets of Langerhans.[49] If these *GLUT2* transporters are deleted from astrocytes in mice, the brain no longer reacts to changes in blood glucose levels by adjusting the functions of the liver and pancreas.[23] Sensory astrocytes, by monitoring levels of blood-borne nutrients, may contribute to the control of feeding behavior by the brain.[8,48]

In another circumventricular organ called the subfornical organ, specialized astrocytes have been found to have another sensory role. These astrocytes have a specialized sodium channel in their membranes that reacts to high levels of sodium. Also, these specialized astrocytes are adjacent to neurons that regulate drinking behavior and regulate their activity. If the astrocyte sodium channel is deleted in genetically modified mice, such mice fail to become thirsty when infused with saline solutions.[42]

Astrocytes and neuronal function

Astrocytes appear to affect neuronal function in a variety of ways. One way is by transfer of nutrients from astrocytes to neurons. It has been estimated that 75% of the glucose entering the brain is first metabolized to *lactate* by astrocytes; the *lactate* is then transferred to neurons as a fuel molecule.[16] Also, astrocytes take up the *glutamate* released from neurons as a neurotransmitter, transform the *glutamate* into *glutamine*, and export the *glutamine* back to neurons so that they can change it back into more *glutamate*. Abnormalities in the glutamate-glutamine cycle between neurons and astrocytes may partly underlie the abnormal firing of neurons seen in epilepsy.[11] Also, pubertal changes in astrocyte glutamate metabolism appear to be a major stimulus to hypothalamic neurons that regulate the release of reproductive hormones from the pituitary, and thus may underlie the onset of reproductive function at puberty.[35]

Astrocytes possess their own specific types of glutamate transporter proteins that allow them to take up *glutamate* from the extracellular space. One such protein (*GLAST*) appears to require *GFAP* for maintaining its anchorage into the plasma membrane of astrocytes.[37]

Astrocytes also modulate the extracellular concentrations of other substances besides nutrients and neurotransmitters. Metals such as zinc can be released from synapses in the hippocampus and other structures; astrocytes possess a metal-binding protein, *metallothionein*, which can protect neurons from toxic effects of zinc and other metals.[3,28,47]

On a more fundamental level, recent experiments indicate that astrocytes have a major role in stimulating the formation of synapses between neurons. Nerve cells grown *in vitro* in the absence of astrocytes have a normal-appearing anatomy, but fail to make synapses.[4] It appears that astrocytes secrete substances like *thrombospondin* and *cholesterol* that have a powerful effect upon the generation of synaptic proteins. Contact of astrocytes with neurons seems to initiate the production of dendritic spines and may be an important contributor to the plasticity of neuronal form and function seen during learning.[27]

Not all the influences of astrocytes in the CNS are as benign as those listed above. For example, after a localized lesion in the CNS, surrounding astrocytes multiply and flock to the region, forming a "glial scar" composed of so-called reactive astrocytes. These reactive astrocytes possess poorly characterized molecules that prohibit the re-growth of axons through the scar. This is a major reason why damage to the CNS, as after a stroke, is so devastating and long-lasting; while in contrast, glial cells of peripheral nerves (Schwann cells) seem to allow re-growth of regenerating axons much more readily. A better understanding of the inhibitory action of CNS astrocytes upon regeneration could prove to be invaluable in treating brain damage.[14]

One additional brain disorder involving astrocytes is Alzheimer's disease (AD). In AD, extracellular accumulations of the *Aβ peptide* form toxic plaques that harm neurons. One way of avoiding the formation of plaques is via degradation of the *Aβ peptide*; this is regulated by a lipid-binding protein secreted by astrocytes called *Apolipoprotein E*. There are four isoforms of this protein that can be produced by astrocytes; if the *E4 version* is the one inherited in a person, the risk for developing AD is greatly increased relative to other people.[17]

Finally, astrocytes may exhibit a number of features that have previously been thought to be unique to neurons. Astrocytes possess cytoplasmic vesicles that contain neurotransmitters like *glutamate* or neuropeptide Y, and when they release these neurotransmitters, the function of synapses is adjusted.[18,34] Thus, neurons are not the only players in the information-communication game in the brain.

Gomori-positive astrocytes

Traditionally, astrocytes have been divided into two subtypes, based upon morphology: protoplasmic astrocytes of grey matter and fibrous astrocytes of white matter. Research such as that presented above shows that astrocytes in different brain regions function differently and probably exist as many different subsets of astrocytes.[15] One interesting subset of astrocytes is that of the Gomori-positive astrocyte. These peculiar-looking cells received their name because these cells possess many large cytoplasmic granules that stain positively with Gomori's chrome alum hematoxylin stain (Fig. 6.1). Where do these cytoplasmic granules come from?

Electron microscopy has shown that cytoplasmic Gomori-positive (GP) granules are derived from degenerating mitochondria that become engulfed within lysosomes![5] Non-digestible remnants of mitochondria — such as iron-rich heme molecules and copper-containing enzyme subunits — account for the intense staining of Gomori — positive granules. These peculiar astrocytes are particularly abundant in the arcuate nucleus of the hypothalamus, but also can be found in the hippocampus of rats and humans.[30]

What has happened to the mitochondria of these cells, and why does this happen in only a few brain regions?

A number of experiments have shown that the GP phenotype and mitochondrial degeneration appear to result from some type of

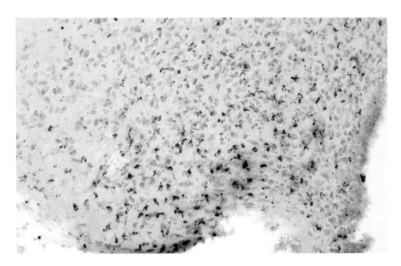

Fig. 6.4. Gomori-positive granules (brown), derived from degenerating mitochondria, are abundant in astrocytes of the arcuate nucleus of the hypothalamus.

oxidative stress.[19] This might make sense, since oxidation of nutrient molecules to provide energy is intense within mitochondria. Also, this mitochondrial degeneration is aging related: GP astrocytes are much more numerous in brains from older rodents than in brains from younger ones. Perhaps toxic effects of molecules diffusing into the arcuate nucleus provoke a gradual, aging-related degeneration of the mitochondria of arcuate astrocytes. But why are neurons not similarly affected?

There are several reasons to believe that mitochondria of astrocytes probably function differently from those of neurons. For example, astrocyte mitochondria possess odd, prismatic cristae, unlike the flattened cristae of most mitochondria (Fig. 1.11). Also, stress-inducing molecules, like the *Aβ peptide* present in Alzheimer's disease, injure astrocyte mitochondria but seem to leave neuronal mitochondria unharmed.[1] Whatever the cause, this damage to astrocyte mitochondria may have implications for aging-related dysfunction of hypothalamic neurons.[40]

A disturbed function of mitochondria has now become a prime suspect in many aspects of aging in general. If repair of mitochondrial DNA is experimentally disrupted in mice, the mice rapidly acquire the symptoms of an accelerated old age: they develop curvature of the spine, hair loss, osteoporosis, anemia, and reduced fertility.[38] So perhaps the mitochondrial degeneration seen in Gomori-positive astrocytes is just a reflection of aging-associated damage that occurs in many organs, including the brain.

2. Oligodendrocytes

The main function of an oligodendrocyte in the brain is to wrap flattened cytoplasmic processes around multiple axons, forming an insulating layer of myelin around each axon. Each cell extends multiple, stubby processes that branch out from the region of the cell nucleus (Fig. 6.5). Curiously, an excellent way to stain oligodendrocytes is to apply a histochemical stain for iron, since iron is very abundant in oligodendrocyte cytoplasm. Iron somehow seems to be essential for the process of myelination of axons; also, the iron-binding

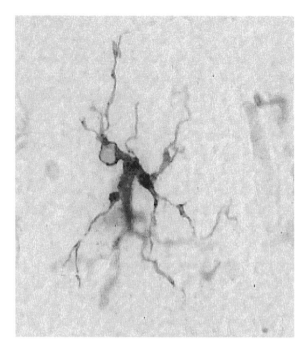

Fig. 6.5.   An example of an oligodendrocyte, with a branching cytoplasm that stains positively for iron (brown) and an oval, unstained nucleus.

protein, *transferrin*, stimulates the differentiation of oligodendrocytes from precursor cells.[31,32] In spite of these data, the specific mechanisms whereby iron affects myelin formation are still not known.

The process of forming myelin from a flattened sheet of cytoplasm and associated membrane involves a number of specialized proteins. An oligodendrocyte-specific cytoskeletal protein called *ermin* appears to be required for the generation of glial cell processes and the deformation of the cytoplasm to produce myelin.[6] In addition, specific membrane proteins are found in the myelin sheath. One of these is called *myelin basic protein*, which helps compact the membranes of myelin by neutralizing the negative charges of membrane phospholipids. The translation of the mRNA for *myelin basic protein* is regulated by an RNA-binding protein called *qkl*. This peculiar abbreviation derives from the name of its gene, *Quaking*; mutations in this gene disrupt the myelination of axons and cause motor abnormalities in affected mice that make them shiver or quake.[51]

About 50% of the protein content of myelin is formed by another protein called *proteolipid protein.* Experimental disruption of this protein in newborn mice also causes severe problems with myelination that lead to death by three weeks of age. In humans, a mutation in this protein causes a rare neurological disorder called Pelizaeus-Merzbacher disease.[10]

Myelination of axons can increase their rate of conductance of an electrical signal (action potential) by as much as 100-fold, so myelination is a vital function of oligodendrocytes.[10] A disturbed myelination causes a number of neurological disorders. One disorder, multiple sclerosis, affects about 1 out of 1000 people in the general population, and causes loss of control of muscles and other symptoms. Multiple sclerosis derives its name from the numerous pin-point CNS lesions (scleroses, or scars) that can be detected in scans of the brain. These lesions result from accumulations of damaged myelin. This disease occurs when helper T-cells that attack myelin cross the blood-brain barrier and stimulate an autoimmune reaction against myelin. The exact reasons why all of this happens are still a matter of considerable debate.[21]

Myelination may not be the only major function of oligodendrocytes. A number of studies have reported that oligodendrocytes may receive synaptic contacts from neurons and may generate neuron-like action potentials in response. Also, an oligodendrocyte may control the rates of transmission of electrical impulses down the axons that it engulfs, possibly coordinating the information flow among bundles of axons.[12]

3. Microglia

Microglia are phagocytic cells that appear to function in a way similar to macrophages (Fig. 6.6).

They do not originate within the brain itself, but arise from mesoderm during development rather than the neural tube and subsequently colonize the brain.[7] They account for about 10% of the

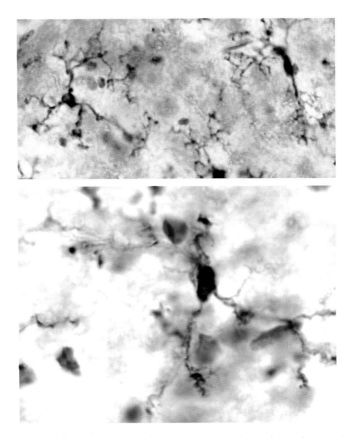

Fig. 6.6. Microglia extend long, delicate processes throughout their environment (neuropil) that stain positively for a protein called CD11b.[24] Photograph courtesy of Dr. Kebreten Manaye, Dept. Physiology, Howard University.

glial cells of the brain.[4] The precise functions of microglia are not well defined; they seem to have some capacity for phagocytosis, but not as great as that of neutrophils or macrophages. Recent imaging studies have shown that, in the normal brain, microglia continually extend and retract their processes, briefly touching synapses every hour or so. However, in conditions of ischemia (diminished blood flow to the brain), microglia become activated and make prolonged contact with synapses. If synapses are damaged by ischemia, microglia seem to have a role in eliminating them.[41] Also, during conditions of general inflammation in the body, microglia produce *monocyte chemoattractant protein* (see Ch. 3), which induces monocytes to migrate into the brain and carry out any phagocytosis of damaged cells or of debris. These steps also seem to be important for the process of clearing away dying cells and debris that form after traumatic brain injury or ischemia.[24]

Secretion of a variety of cytokines such as *interleukin-1* by microglia seems to play a role in the response of the brain to injury or infections.[4,9] One disease that is of particular interest is AIDS (infection by the HIV virus). A frequent occurrence in the final stages of HIV infection is the appearance of dementia and degenerative changes in neurons. These changes do not seem due to direct effects of the virus on neurons, but to infection of microglia by the virus. Microglia appear to respond by secreting some uncharacterized substance that suppresses a process called autophagy in neurons. Autophagy is defined as the digestion of damaged organelles and membranes within lysosomes; it is essential for ridding a cell of abnormal structures. An abnormal regulation of neuronal autophagy by microglia seems to be an important component of the brain damage seen in AIDS patients.[2]

Recent studies suggest that, although microglia throughout the brain have similar morphologies, they may be divisible into two subtypes that vary in their secretion of cytokines, in cell surface markers, and in protective vs. damaging influences upon neurons.[36]

For the last hundred years of neuroscience, glia have been the "Rodney Dangerfield" cells of the brain, commanding little respect from neuroscientists. The data presented above show that this condition of ill-repute for glia will likely not persist much longer.

References

1. Abramov AY, Canevari L, Duchen MR (2004) β-amyloid peptides induce mitochondrial dysfunction and oxidative stress in astrocytes and death of neurons through activation of NADPH oxidase. *J Neurosci* 24: 565–575.
2. Alirezaei M, Kiosses WB, Fox HS (2008) Decreased neuronal autophagy in HIV dementia: a mechanism of indirect neurotoxicity. *Autophagy* 4: 963–966.

3. Aschner M, *et al.* (1998) Induction of astrocyte metallothioneins (MTs) by zinc confers resistance against the acute cytotoxic effects of methylmercury on cell swelling, Na+ uptake, and K+ release. *Brain Res* 813: 254–261.

4. Barres BA (2008) The mystery and magic of glia: a perspective on their roles in health and disease. *Neuron* 60: 430–440.

5. Brawer JR, *et al.* (1994) Composition of Gomori-positive inclusions in astrocytes of the hypothalamic arcuate nucleus. *Anat Rec* 240: 407–415.

6. Brockschnieder D, *et al.* (2006) Ermin, a myelinating oligodendrocyte-specific protein that regulates cell morphology. *J Neurosci* 26: 757–762.

7. Chan WY, Kohsaka S, Rezaie P (2007) The origin and cell lineage of microglia: new concepts. *Brain Res Rev* 53: 344–354.

8. Cheunsuang O, Morris R (2005) Astrocytes in the arcuate nucleus and median eminence that take up a fluorescent dye from the circulation express leptin receptors and neuropeptide Y Y1 receptors. *Glia* 52: 228–233.

9. D'mello C, Le T, Swain MG (2009) Cerebral microglia recruit monocytes into the brain in response to tumor necrosis factor α signaling during peripheral organ inflammation. *J Neurosci* 29: 2089–2102.

10. Dimou L, *et al.* (1999) Dysmyelination in mice and the proteolipid protein gene family. In *The Functional Roles of Glial Cells in Health and Disease* (eds.) M Matsas, M Tsacopoulos, Plenum Publ.: NY, 1999, pp. 261–271.

11. Eid T, *et al.* (2008) Glutamate and astrocytes — key players in human mesial temporal lobe epilepsy? *Epilepsia* 49(S2): 42–52.

12. Fields RD (2008) Olidogendrocytes changing the rules: action potentials in glia and oligodendrocytes controlling action potentials. *Neuroscientist* 14: 540–543.

13. Herculano-Housel S, Lent R (2005) Isotropic fractionator: a simple, rapid method for the quantification of total cell and neuron numbers in the brain. J *Neurosci* 25: 2518–2521.

14. Hirsch S, Bähr M (1999) Growth promoting and inhibitory effects of glial cells in the mammalian nervous system. In *The Functional Roles of Glial Cells in Health and Disease* (eds.) M Matsas, M Tsacopoulos, Plenum Publ.: NY, 1999, pp. 199–203.

15. Hochstim C, *et al.* (2008) Identification of positionally distinct astro-cytes subtypes whose identities are specified by a homeodomain code. *Cell* 133: 510–522.

16. Hyder F, *et al.* (2006) Neuronal-glial glucose oxidation and glutamatergic-GABAergic function. *J Cerebral Blood Flow & Metab* 26: 865–877.

17. Jiang Q, *et al.* (2008) ApoE promotes the proteolytic degradation of Aβ. *Neuron* 58: 881–893.

18. Jourdain P, *et al.* (2007) Glutamate exocytosis from astrocytes controls synaptic strength. *Nat Neurosci* 10: 331–339.

19. Justino L, Welner SA, Tannenbaum GS, Schipper HM (1997) Long-term effects of cysteamine on cognitive and locomotor behavior in rats: relationship to hippocampal glial pathology and somatostatin levels. *Brain Res* 761: 127–134.

20. Koehler RC, Roman RJ, Harder DR (2009) Astrocytes and the regula-tion of cerebral blood flow. *Trends Neurosci* 32: 160–169.

21. Korn T (2008) Pathophysiology of multiple sclerosis. *J Neurol* 255(S6): 2–6.

22. Lee SW, *et al.* (2003) SSeCKS regulates angiogenesis and tight junction formation in blood-brain barrier. *Nat Med* 9: 900–906.

23. Marty N, *et al.* (2005) Regulation of glucagon secretion by glucose transporter type 2 (glut2) and astrocyte-dependent glucose sensors. *J Clin Invest* 115: 3545–3553.

24. Matsumoto H, *et al.* (2007) Antibodies to CD11b, CD68, and lectin label neutrophils rather than microglia in traumatic and ischemic brain lesions. J Neurosci Res 85: 994–1009.

25. Menon RS (2001) Imaging function in the working brain with fMRI. *Curr Opin Neurobiol* 11: 630–636.

26. Muroyama Y, *et al.* (2005) Specification of astrocytes by bHLH protein SCL in a restricted region of the neural tube. *Nature* 438: 360–363.

27. Nishida H, Okabe S (2007) Direct astrocytic contacts regulate local maturation of dendritic spines. *J Neurosci* 27: 331–340.

28. Nolte C, *et al.* (2004) ZnT-1 expression in astroglial cells protects against zinc toxicity and slows the accumulation of intracellular zinc. *Glia* 48: 145–155.

29. Oberheim NA, *et al.* (2009) Uniquely hominid features of adult human astrocytes. *J Neuroscience* 29: 3276–3287.

30. Ohm TG, Jung E, Schnecko A (1992) A subpopulation of hippocampal glial cells specific for the zinc-containing mossy fibre zone in man. *Neuroscience Letters* 45: 181–184.

31. Ortiz E, *et al.* (2004) Effect of manipulation of iron storage, transport, or availability on myelin composition and brain iron content in three different animal models. *J Neurosci Res* 77: 681–689.

32. Paez PM, *et al.* (2004) Apotransferrin promotes the differentiation of two oligodendroglial cell lines. *Glia* 46: 207–217.

33. Peruzzo B, *et al.* (2000) A second look at the barriers of the medial basal hypothalamus. *Exp Brain Res* 132: 10–26.

34. Ramamoorthy P, Whim MD (2008) Trafficking and fusion of neuropeptide Y-containing dense core granules in astrocytes. *J Neurosci* 28: 13815–13827.

35. Roth CL, *et al.* (2006) Quantitative proteomics identifies a change in glial glutamate metabolism at the time of female puberty. *Mol Cell Endocrinology* 254–255: 51–59.

36. Sawada M (2009) Neuroprotective and toxic changes in microglia in neurodegenerative disease. *Parkinsonism Relat Disord* 15(S1): 539–541.

37. Sullivan SM, *et al.* (2007) Cytoskeletal anchoring of GLAST determines susceptibility to brain damage: an identified role for GFAP. *J Biol Chem* 282: 29414–29423.

38. Trifunovic A, *et al.* (2004) Premature ageing in mice expressing defective mitochondrial DNA polymerase. *Nature* 429: 417–423.

39. Vannucci SJ, Maher F, Simpson IM (1997) Glucose transporter proteins in brain: delivery of glucose to neurons and glia. *Glia* 21: 2–21.

40. Voloboueva LA, *et al.* (2007) Inhibition of mitochondrial function in astrocytes: implications for neuroprotection. *J Neurochem* 102: 1383–1394.

41. Wake H, *et al.* (2009) Resting microglia directly monitor the functional state of synapses *in vivo* and determine the fate of ischemic terminals. *J Neurosci* 29: 3974–3980.

42. Watanabe E, *et al.* (2006) Sodium-level-sensitive sodium channel Na_x is expressed in glial laminate processes in the sensory circumventricular organs. *Am J Physiol Regul Integr Comp Physiol* 290: R568–R576.

43. Winship IR, Plaa N, Murphy TH (2007) Rapid astrocyte calcium signals correlate with neuronal activity and onset of the hemodynamic response *in vivo*. *J Neurosci* 27: 6268–6272.

44. Wolburg H, *et al.* (2009) Brain endothelial cells and the glio-vascular complex. *Cell Tiss Res* 335: 75–96.

45. Yan J-W, Suder P, Silberring J, Lubec G (2005) Proteome analysis of mouse primary astrocytes. *Neurochem Int* 47: 159–172.

46. Young JK, Wang C (1990) Glucose transporter immunoreactivity in the hypothalamus and area postrema. *Brain Res Bull* 24: 525–528.

47. Young JK, Garvey JS, Huang PC (1991) Glial immunoreactivity for metallothionein in the rat brain. *Glia* 4: 602–610.

48. Young JK (2002) Anatomical association between specialized astrocytes and leptin-sensitive neurons. *J Anat* 201: 85–90.

49. Young JK, McKenzie JC (2004) GLUT2 immunoreactivity in Gomori-positive astrocytes of the hypothalamus. *J Histochem Cytochem* 52: 1431–1436.

50. Young JK (2006) Astrocytes and glucose sensing by the brain. in *Recent Research Developments in Molecular and Cellular Biology.* ed. SG Pandalai, Vol. 6, pp. 1–15. Research Signpost, Kerala, India.

51. Zearfoss NR, Farley BM, Ryder SP (2008) Post-transcriptional regulation of myelin formation.

52. Zeller K, Vogel J, Kuschinsky W (1996) Postnatal distribution of glut1 glucose transporter and relative capillary density in blood-brain barrier structures and circumventricular organs during development. *Brain Res Dev Brain Res* 91: 200–208.

Chapter 7

HOW ARE THE NUMBERS OF CELLS IN AN ORGAN REGULATED?

In order to function properly and obtain the precise size required for that function, every organ of our bodies must acquire the exact number of cells needed. This is one of the most basic, and yet most poorly understood, principles of anatomy and physiology.

For many organs, some type of instructions seems to precisely limit the number of cells and the size of an organ. This can be demonstrated experimentally: if multiple spleens are transplanted into a newborn mouse, they don't all grow to their normal size, but eventually acquire the normal mass of a single spleen by adulthood. Somehow, the body exerts some extrinsic control over the size of the spleen so that it does not expand inappropriately. In another example, a liver may be stimulated to increase in size by administration of phenobarbital, but afterwards, the liver will shrink back to normal size via drastic increases in death rates of liver cells.[4]

The very dimensions of our bodies also seem tightly regulated. For example, we generally take for granted that our right limbs are almost exactly the same size as our left limbs, but the mechanisms that control the size of these organs are in fact not well understood. The right humerus of most humans is slightly longer (5 mm) than the left humerus, probably because most of us are right-handed and the muscles of the right extremity twist and strain the bones of our right arm more and stimulate mechanisms for compensatory bone growth. The left and right tibial bones, however, rarely differ from each other in length by more than 1–2 mm; this is very fortunate, since even minute disparities in leg bone length would lead to impaired walking.[2] From infancy to adulthood, these limb bones undergo expansions of length of at least 250%, but eventually show right-left length differences that are very minute. How are the sizes of these organs so closely regulated?

It could be argued that right-left symmetry is simply due to the fact that the cells of both extremities are exposed to the same amount of circulating growth hormone and respond identically. However, data from examples of specific genetic syndromes that are inherited unilaterally refute this hypothesis. For example, rare individuals may experience genetic damage in only a portion of cells in the embryo that causes a unilateral loss of functional elastic fibers (Marfan's syndrome); in these cases, the limbs on the affected side may be 2–3 cm longer than on the normal side.[11] Since elastic fibers are not even present in bone, this bone elongation likely results from a diminished restraint of bone growth by surrounding connective tissues in the affected side. This hypothesis is supported by the fact that children with unequal leg length, due to injury or other causes, can be helped by a circumferential incision of the periosteum covering the shortened leg bone, which releases it from tension and allows a further growth.[24] These types of data clarify the issues involved in controlling limb size, but they by no means provide all the answers for this problem.

The number of cells in an organ can also show differences related to species, age, and nutritional state. For example, the cardiac muscle cells of a mouse greatly resemble those of humans in size and appearance, but a human heart weighs almost one thousand times more than a mouse heart and contains at least 300-fold more cells.[16] How are embryos controlled so that the size of the heart is appropriate for the organism? If the heart were to become only twice its normal size, this would have a devastating effect upon development. No certain answers to this question are known.

Almost all organs experience loss of cells due to damage and birth of new cells to replace them. Some changes in cell number, however, are more drastic than others. For example, during pregnancy, the weight of the uterus increases from 30 grams to over 700 grams.[12] Much of this growth is due to vastly increased numbers of smooth muscle cells within the muscular layer of the uterus (myometrium). This response to pregnancy depends upon cellular reactions to estrogen and to stretch, which provoke the appearance of *Insulin-like Growth Factor 1* (*IGF1*) within the myometrium. Smooth muscle cells proliferate greatly in response to *IGF1*. Following the birth of a baby, however, the uterus rapidly decreases in size and reverts to normal by four weeks postpartum. This decrease requires massive amounts of cell death within the uterus.[12,22] The death of extra cells, in this case at least, explains how the size of an organ is controlled.

The term "cell death" should actually not be applied loosely here. Obviously, cells can die due to severe injury: holding my finger over a lit candle will produce lots of cell death. But this type of cell death, due to externally applied injury, is termed necrosis and results in the bursting of the cell membrane, a release of a cell's contents into its environment, and the recruitment of white blood cells to clean up the mess (inflammation). In contrast, mechanisms of cell death normally employed to control cell number involve a form of cell death termed apoptosis (from the Greek *apo* — "high above" and *ptosis* — "to fall from", in analogy with leaves programmed to die and fall away from a tree). In apoptosis, intracellular mechanisms damage cell constituents, but the cell does not rupture or provoke inflammation. Apoptosis is not a response to external injury, but is one mechanism among many that are used to control cell number in an organ by balancing the ratio of cell birth to cell death.

The ovary is another example of an organ that experiences dramatic changes in cell number with age. An ovary of a 19-year-old woman contains about 300,000 healthy oocytes within ovarian follicles; by age 50, the number of oocytes will decline to only 1500.[8] This is not due to aging of the egg cells themselves, but seems to be due to selective apoptosis of the granulosa cells that surround each given oocyte. The factors that select which ovarian follicles will degenerate and which will mature to ovulation are still mysterious.[18]

The liver is an organ that can show remarkable changes in cell number and has an amazing capacity for regeneration. In rats, after surgical removal of 70% of the liver, the remaining liver remnant can regenerate and completely replace the normal volume of the liver within a week! Two fundamental questions can be asked about this fascinating ability: 1) how do the remaining liver cells "know" that much of the liver has been removed, so that they can multiply in response, and 2) how do the liver cells "know" when to *stop* multiplying so that they do not completely fill up the entire abdominal cavity?

One answer to the first question now seems to be at hand. When much of the liver is removed, the blood vessels entering the liver remnant are spared and continue to deliver about the same volume of blood to the liver remnant as had been flowing previously in the entire liver. This results in a 3–10 fold increase in blood flow within liver sinusoidal vessels. The endothelial cells lining these vessels experience a shear stress from this increased blood flow and react by secreting growth factors such as *hepatocyte growth factor*, which stimulate cell division of hepatocytes.[21] Presumably, as the liver enlarges and its capacity to contain blood grows, the stress upon blood vessels abates and the production of growth factors subsides. Is this the only explanation for why the liver does not continue to grow? The answer to this second question will be discussed below as we explore mechanisms that control cell number.

A final organ that experiences drastic shifts in cell number is the developing brain. Almost half of the neurons that develop from the neural tube will die prior to the birth of a rat.[3] One theory for this massive cell death is that it appears to be a mechanism for disposing of neurons that fail to make functional synaptic connections with targets. This theory has a lot of experimental support. For example, sensory neurons of developing sensory ganglia send neurites out towards the skin, where they contact cells that produce *Nerve Growth Factor* (*NGF*). If the skin of an embryo is experimentally induced to make extra *NGF*, this growth factor supports the continued life of many more sensory neurons, so that the trigeminal ganglia of altered mice eventually contain 57,000 neurons rather than the normal 26,000.[1] Thus, functional contact with a target is required for a neuron to avoid death, and if the growth-promoting effects of a contact are enhanced, cell death is decreased.

Cell death occurs in the adult brain as well as in the developing nervous system. In an uninjured brain, apoptosis mainly occurs in regions where new neurons are produced (dentate gyrus of the hippocampus, olfactory bulb).[3] However, various types of brain injury, such as cutting of axons, can lead to neuronal apoptosis in any brain region. Curiously, experimentally induced apoptosis in the cerebral cortex can cause the spontaneous appearance of newborn, dividing neurons that occupy the places of their dead predecessors and which appear to try to provide substitutes for missing axons and synapses.[17]

Neuronal apoptosis can also result from exposure to excessive levels of the excitatory neurotransmitter, *glutamate*, which has therefore been termed an "excitotoxin." *Glutamate* excitotoxicity may underlie brain damage in a number of neurological disorders (Huntington's disease, Alzheimer's disease); also, transient ischemia (impaired blood flow) causes cell death in specific areas of the hippocampus that may be due to a failure of astrocytes to remove damaging levels of *glutamate* from the area.[20]

Exposure of neurons to glutamate causes a specific sequence of events: 1) at one hour after *glutamate* exposure, numbers of autophagic vesicles increase, 2) at eight hours after *glutamate* exposure, many cellular organelles have become surrounded by autophagic vesicles and have been destroyed, and 3) at sixteen hours after *glutamate* exposure, the cell nucleus becomes shrunken and darkened and the cell is well on its way to death.[23] These events can be provoked experimentally by intracranial infusions of *glutamate*. They can also be more easily observed in the arcuate nucleus of the hypothalamus. Since this brain region has unusually permeable capillaries (see Ch. 6), an intravenous infusion of *glutamate* easily penetrates into the arcuate nucleus and provokes apoptosis of neurons there (Fig. 7.1). When this was discovered, a recommendation was made to forbid the use of *monosodium glutamate* in baby food, since infants do not metabolize *glutamate* as rapidly as adults and thus could conceivably show neurotoxic levels of *glutamate* in the bloodstream after a large meal that could conceivably damage the hypothalamus.[14]

Why do these events happen, and what molecules regulate autophagy and apoptosis?

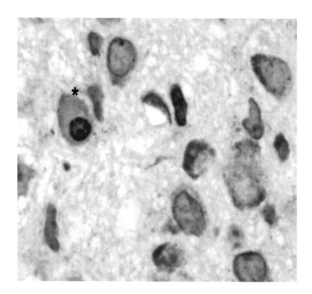

Fig. 7.1. High magnification view of the arcuate nucleus of the hypothalamus, showing a neuron (*) with a shrunken, condensed nucleus that is undergoing apoptosis in response to glutamate. Nearby, unaffected neurons have large, pale nuclei.

1. Events and Mechanisms of Apoptosis

The first step in glutamate-induced apoptosis of neurons is an increase in autophagic vacuoles. What are these? Autophagic vacuoles begin their existence when many small vesicles, termed preautophagosomes, coalesce in a small region of the cytoplasm.[5] This event is governed by as many as 20 proteins coded for by so-called *AuTophagy related Genes (ATGs)*.[13] These *Atg* proteins bind to each other and also to lipid molecules to create a flattened, saucer-shaped membranous sac that rather resembles one stack out of the many forming the Golgi apparatus (see Fig. 1.8). This sac moves towards a damaged organelle — for example, a mitochondrion — and folds around it until it completely surrounds and engulfs the organelle in a spherical, two-layered bubble.

What causes this to happen? In *glutamate*-induced apoptosis, a critical event is a *glutamate*-stimulated influx of calcium into the cell. One cellular storage area for calcium is the mitochondrion, which can absorb excess calcium and store it as dense granules within the mitochondrial matrix. However, excessive amounts of calcium disturb the balance of ions within a mitochondrion and cause a pore called the mitochondrial transition pore to form in the inner mitochondrial membrane. This allows ions to flow incorrectly within a mitochondrion and cause a depolarization of the mitochondrion.[14] Calcium is not the only stimulus to provoke this; starvation will also deplete cellular ATP and causes a five-fold increase in the number of depolarized mitochondria within a cell, amounting to 1% of all mitochondria. Thus, calcium and diminished ATP can both promote mitochondrial abnormalities.[5,15]

Somehow, a disabled, depolarized mitochondrion stimulates the formation of an autophagic vacuole and its own subsequent engulfment.[15] This spells doom for the mitochondrion, for it signifies an upcoming close encounter with a lysosome.

Lysosomes are spherical vesicles filled with 20–30 types of hydrolytic enzymes (including a category called *cathepsins*) that are activated at an acidic pH. These enzymes have a powerful ability to digest lipids and proteins into their component molecules. About 50% of the integral membrane proteins of a lysosome belong to the so-called *LAMP* (*Lysosomal Associated Membrane Protein*) family. Each of the four types of *LAMP* proteins bears a specific sequence of five amino acids on its C-terminal end that directs the protein to be inserted into a lysosomal membrane.

Several aspects of lysosomal function would appear to be problematic. They contain highly destructive, dangerous enzymes. This presents a challenge to the cell: if you have made a universal solvent, what do you store it in? How can the membranes enclosing a lysosome be kept safe from digestion? The answer to this question is not completely known, but it is certain that the *LAMP* proteins become highly glycosylated (decorated with carbohydrate molecules) when they pass through the rER and Golgi apparatus on their way to lysosomes. This may protect them from the destructive effects of the enzymes they enclose.

Another challenge for a cell is to make sure that dangerous lysosomal enzymes are sent only to lysosomes and not to other locations in the cell. This is accomplished by adding a molecule of *mannose-6-phosphate* to each enzyme as it is synthesized in the rER. Certain patches of rER membrane possess mannose-6-phosphate receptor proteins that bind newly synthesized lysosomal enzymes. When this binding event occurs, the patches bud off from the rER as vesicles that travel only to other organelles called late endosomes. This movement involves *AP1 type adaptor complex proteins* that bind to the *mannose-6-phosphate* receptor proteins in vesicle membranes and which guide the vesicles to fuse only with late endosomes (see Ch. 2 for a discussion of adaptor complex proteins).

Late endosomes are vesicles that possess a somewhat acidic internal pH, because they transport H^+ ions into their interior. This acidic pH causes the lysosomal enzymes to pop off of the *mannose-6-phosphate* receptor proteins, and the enzymes are freed to float around within the fluid filling a late endosome. How these enzymes are subsequently passed into lysosomes is still not clear.[25]

The *LAMP* proteins that help form the structure of lysosomes also have another function: they appear to be necessary for the fusion of an autophagosome with a lysosome. If *LAMP2* is experimentally deleted from a cell, massive amounts of autophagosomes accumulate and digestion of cell organelles is stopped.[10] In humans, a rare mutation of *LAMP2* has similar effects and causes a disorder called Danon disease, in which there are extensive abnormalities in muscle and nerve cells.[19] The careful control of lysosomal fusion with other membranes by *LAMP* proteins ensures that lysosomes do not fuse with normal cell components and digest them. When lysosomes do fuse with an autophagosome, its contents are digested into smaller molecules that are transported out of the lysosome via amino acid transporter proteins and other membrane molecules.

Lysosomes are not the only organelles devoted to the destruction of biological molecules. Another similar-appearing organelle, the peroxisome, also functions in cellular destruction. Each small, spherical peroxisome contains about 50 enzymes. One such enzyme, *catalase*, converts potentially dangerous molecules of hydrogen peroxide to water. Other peroxisomal enzymes are devoted to other functions such as lipid and amino acid oxidation. All of these enzymes reach the peroxisome from their sites of synthesis in the cytoplasm by binding to a peroxisomal membrane protein called *peroxisomal receptor 1*; this membrane protein binds to a serine-lysine-leucine signal of amino acids found at one end of peroxisomal enzymes. Abnormalities in the peroxisomal receptor cause a number of severe disorders such as Zellweger syndrome, in which a lack of normal peroxisomal function damages the brain, liver, and kidneys.

All of the above may give you the impression that autophagocytosis is an abnormal, pathological process. It is not. Autophagocytosis provides a cell with a way to digest the components of damaged organelles within lysosomes and to reutilize the molecules transported out of lysosomes for new purposes. All organelles, except for the nucleus, are recycled in this way; the half-life of a mitochondrion in a cell typically amounts to only 10–25 days.[15] Autophagocytosis also helps a cell adapt to changing nutritional circumstances. The nutrient state of a cell is monitored by a protein called *TOR* (see Ch. 4 for more details about *TOR*). *TOR* normally depresses autophagocytosis by interacting with a number of *Atg* proteins; during starvation, this *TOR*-mediated depression ceases and a cell begins to cannibalize its own organelles for spare parts.[5]

Organelles are not the only structures that are digested in lysosomes. Lysosomes also destroy portions of the cell membrane, in particular, they destroy cell membrane receptors for signaling molecules like hormones. This provides cells with a mechanism for "turning off" the response to a hormone. This mechanism, termed receptor-mediated endocytosis, causes the receptors for a hormone to cluster in one place on the cell membrane. This cell membrane then invaginates to form a small endocytic vesicle that is shaped on its cytoplasmic surface by a protein called *clathrin*. These vesicles migrate to larger organelles called early endosomes, and then finally fuse with late endosomes. Fusion of late endosomes with lysosomes destroys cell membrane receptors for hormones so that the cell can resume a more normal function.

While extensive autophagocytosis can damage much of a cell, it is not always sufficient to cause apoptosis. A major stimulus for apoptosis is the release from mitochondria of a component of the electron transport chain, a molecule called *cytochrome C*. *Cytochrome C* binds with a protein called *Apaf-1* (*apoptotic peptidase activating factor*) to form a small structure called an apoptosome; this structure goes on to cleave an inactive protein (*procaspase*) into a smaller, active enzyme called *caspase-9*.[14] This is the start of the process for cell death.

Caspases are essential components of the program for cell death. These constitute a family of 11 proteolytic enzymes that have a Cysteine at their active site and which cleave proteins at ASPartate residues (hence the name, *caspase*). *Caspase-3*, when cleaved and activated by *caspase-9*, appears to cause much of the dissolution of structural proteins of the cytoplasm and activates still other *caspases*. As a result, the nuclear envelope becomes permeable, enzymes digest DNA, and the nucleus becomes shrunken and condensed. Finally, the cell breaks apart into apoptotic vesicles that are phagocytized by macrophages.[14,18,23]

Apoptosis is not always inevitable in a damaged cell. Some cellular proteins can prevent or postpone apoptosis. A family of about 25 proteins is known to modulate apoptosis. These are *Bcl proteins*, which derive their name from B-cell Lymphomas, cancers with runaway cell division and reduced cell death in which they were first described. *Bcl* proteins prevent *cytochrome C* from activating *caspase-9*.[14] Another anti-apoptotic protein called *survivin*, which does not belong to the *Bcl* family, appears to play a role in maintaining the proper numbers of cells in the heart.[16] Finally, apoptosis is not always initiated by mitochondrial damage. A number of proteins, such as *Tumor Necrosis Factor* or *Fas ligand*, can initiate apoptosis and *caspase* activation by binding to cell surface receptors.

What happens to a cell when it undergoes apoptosis? Such a cell is easily recognized by macrophages, because a damaged cell has an abnormal distribution of molecules in its cell membrane. One such molecule is *phosphatidylserine* (*PS*). *PS* is normally present only on the cytoplasmic surface of a cell membrane; when a cell undergoes apoptosis, enzymes called *scramblases* cause *PS* to appear on the external surface of a cell. This signals a macrophage to engulf and phagocytize dying cells[9] (Fig. 7.2).

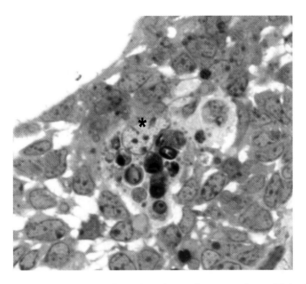

Fig. 7.2. A pale-staining, bean-shaped nucleus of a macrophage (*) is visible in this atretic ovarian follicle. It is surrounding by dark-staining, phagocytized debris originating from apoptotic cells.

2. Other Mechanisms for the Control of Cell Number

In addition to enhancing cell death, organs can control their size by inhibiting cell division. One potent inhibitor of cell division, which can be demonstrated well be observing cells *in vitro*, is the phenomenon of contact inhibition. This refers to the fact that when normal cells in a dish multiply so rapidly that they begin to all touch each other, they automatically cease dividing. One disturbing feature of cancer cells is that they have lost this contact inhibition, and will continue to grow into piles of cells regardless of how many neighbors that they touch. But what mechanism explains contact inhibition?

Recent work conducted in *Drosophila* and in mammals has identified a universal size-control mechanism that controls organ size and is related to contact inhibition of cell multiplication. In mammals, the main components of this mechanism are several enzymes (*kinases* that phosphorylate proteins) and a protein called *Yap* that binds to a transcription regulating protein and which dictates the transcription or repression of many genes reguired for cell division.

The kinases involved in this pathway are called *MST1* (for *Mammalian STerile-like kinase*, named after a yeast homologue that makes yeast cultures sterile) and *LATS* (for *LArge Tumor Suppressor*). When activated, *MST1* phosphorylates *LATS*, and activates it. *LATS* then phosphorylates the *Yap* protein. This prevents the *Yap* protein from migrating to the nucleus, its site of action; thus, *MST1* and *LATS* combine to prevent *Yap* from doing its job.

What is the job of *Yap*? The name for the *Yap* protein originally came from its association with another kinase called *Yes*; thus *Yap* stands for *Yes Associated Protein*. However, as it turns out, *Yap* has far more to do than simply associate with a cytoplasmic kinase. When de-phosphorylated and active, *Yap* moves to the cell nucleus and settles down on the DNA in association with a poorly characterized transcription regulating protein. When bound to DNA, it activates the transcription of about 20 proteins required for cell proliferation (*polo-like kinase*, 3 types of growth factors, *c-myc*, *Sox4*, etc.). It also depresses the transcription for 16 proteins that keep cells quiescent and tightly attached to each other (eg., cell junction proteins).[6,27] *Yap* decreases apoptosis by promoting the transcription of a member of the *Bcl2* family of proteins called *MCL1*, and also increases the production of *survivin*, a non-*Bcl2*, anti-apoptotic protein.[6] Thus, *Yap* is a potent activator of cell proliferation and migration and a suppressor of cell death. When *Yap* is experimentally activated in the liver of a mouse, the liver will grow to four times its normal size and start to develop tumors![6]

How is *Yap* regulated? The answers to this question are not certain, but it appears that some candidate cell surface receptors may be involved and may regulate the activity of the kinases that control *Yap*.[27] Thus, one of the basic regulatory pathways for the control of organ size by contact inhibition is just now becoming understood. This new knowledge is likely to have important consequences, not only for the control of organ growth, but also for an understanding of tumor growth.

Direct cell-cell contact is probably not the only means by which the size of an anatomical structure can be controlled. Cells may also secrete molecules into their environment that inhibit the growth of

Chapter 8

HOW DO SENSORY CELLS FUNCTION?

Almost any cell in the body could be termed a sensory cell. Most cells show some irritability (reactivity to their surroundings), and many cells like bone cells, cartilage cells or kidney cells have sensory cilia that keep them appraised of changes in their fluid-filled environments. However, it is nevertheless fair to say that certain epithelial and nerve cells show a special sensitivity to events in their environment and communicate their reactions to the CNS. These are the subjects of this chapter.

1. Types of Sensory Receptors

Before any specific sensor can be discussed, it is important to recognize that the function of any sensor is dependent upon one of several basic types of receptor proteins embedded in the cell membrane. These include 1) *G-protein linked receptors*, 2) *ligand-gated ion channels*, and 3) *cytokine receptors*. The molecules that bind to these receptors are termed ligands.

G-protein linked receptors are typically integral membrane proteins with 7 intramembrane domains. These receptors, when activated, associate with adjacent proteins called *G-proteins*, which alter intracellular levels of intracellular messengers such as *cyclic AMP* or *cyclic GMP* (Fig. 8.1). This leads to an altered activity of protein kinases and a phosphorylation of specific proteins that alter the function of a cell. *Protein kinase A*, when activated by *cAMP*, can also move to the nucleus and activate the transcription regulating protein *CREB* (see Ch. 3). *G-protein coupled receptors*, as we shall see, are important for senses such as vision, olfaction, and taste.

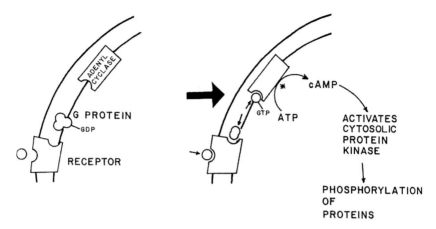

Fig. 8.1.   Diagram of the function of a *G-protein coupled receptor*. When the receptor binds a ligand, it associates with a tripartite *G protein* that has α, β, and γ subunits. The α subunit of the *G protein* acquires a molecule of GTP and dissociates from the other subunits. This α subunit then associates with a membrane-bound enzyme, *adenyl cyclase*, which generates *cyclic AMP* from ATP. *Cyclic AMP*, in turn, activates *protein kinase A*.

Other types of receptor proteins are associated with membrane proteins that form a pore in the cell membrane that can be opened to allow the passage of a specific ion. These types of receptors are important for sensations of touch. Finally, *cytokine* receptors respond to a ligand by activating the *Janus kinase-STAT* pathway, which results in the movement of a transcription regulating protein to the nucleus. These receptors are important for detecting specific types of circulating hormones, and will be discussed in more detail below.

2. Sensors of the Skin

Nerve endings found just beneath the skin represent the terminal portions of the sensory cells with functions that are most basic for us. These nerve endings mostly originate from neurons located in the dorsal root ganglia, next to the spinal cord (see Fig. 2.7 in Ch. 2), or in sensory ganglia for the cranial nerves of the face like the trigeminal ganglion.

Sensations of touch

About 90% of the sensory fibers innervating the skin respond mainly to mechanical stimulation, i.e., touch or pressure. These nerves do not look especially different from other sensory nerves, so what makes them specialized to react to touch?

The touch-reactive sensory neurons of the dorsal root ganglia make a specialized protein called *SLP-3* (named *Stomatin Like Protein-3*, because of its similarity to a protein, *stomatin*, found in red blood cell membranes). This protein forms part of a sodium channel located in the plasma membrane of touch-sensitive axons. If it is experimentally deleted in mice, recordings from the cutaneous axons of mice show that they have lost sensitivity to rapid taps on the skin, but not to other stimuli. It was logical to conclude from these data that the *SLP-3* protein is important for the sense of touch and that mice deficient in *SLP-3* would have an impaired tactile ability, but how was this to be proven? It was impossible to ask the mice if they had trouble detecting objects by touch.

This problem was solved by simply placing the mice in a light-proof, darkened box with a special feature: investigators glued a small piece of sandpaper to the floor, with the rough side of the paper face up. Then, by using a beam of invisible, infrared light and a detector, the researchers analyzed where the mice walked in the box. Normal mice would pause frequently at the sandpaper to touch and feel it. Mice lacking the *SLP-3* protein would ignore the sandpaper and clearly had a sensory deficit. This was pretty good evidence for how the *SLP-3* protein contributes to touch sensation. It appears that stretching of the plasma membrane somehow changes the configuration of this protein and opens the sodium channel to depolarize the axon.[36]

Sensation of temperature

Other cutaneous nerve endings, which still have much the same morphology as other nerve endings, are specialized to react to specific temperatures. Curiously, the molecular mechanisms for temperature perception were first identified via the use of food flavorings. A molecule found in spicy foods (e.g., hot chilis) called *capsaicin* has long been known to provoke a feeling of heat in the mouth. This molecule belongs to a wide variety of plant compounds called *vanillins*. Plants evolved to make these types of compounds as a defensive measure: animals that ate them encountered unpleasant taste sensations and were less likely to eat that plant again!

Application of *capsaicin* to extracts of neural tissue allowed the identification of the protein that bound *capsaicin* to provoke a sensation of heat. This protein was initially termed the *vanillin receptor 1* (*VR1*); since then, however, it has become clear that this protein is only one of a family of related proteins that are now called *Transient Receptor Potential (TRP) receptors*. There are 33 such proteins in humans, divisible into six subfamilies. Each protein is an integral membrane protein that associates with 3 other family members to comprise a calcium channel in the cell membrane.[11]

The *VR1* protein appears to react to an elevation in temperature by allowing the entry of calcium into a sensory axon, thus sending a signal of heat to the brain. *Capsaicin* binds to this protein and forces the channel to stay open even in the absence of heat. Another *TRP* protein, sensitive to the cool-seeming taste of *menthol* in candies, opens in response to cool temperatures and mediates the sensation of cold. Some *TRPs* bind to molecules released from connective tissue mast cells like *histamine* or *bradykinin*. These mediate the painful sensations associated with inflammation, and also react to noxious stimuli like the *acrolein* molecules of tear gas.[1]

Nerve cells are not the only cells that possess *TRP* proteins. One type of protein, found in the cell membranes of smooth muscle cells, reacts to stretch and appears to be important for the control of blood vessel reactions to high blood pressure.[11]

Painful sensations

Stimuli that evoke the sensation of pain are generally those which are intense enough to cause tissue damage, e.g., a crushing stimulus as opposed to a tapping stimulus, or a burning stimulus as opposed to a warm stimulus. It is philosophically and scientifically difficult for a rational human being to define the boundaries between harmless and painful stimuli. How do simple nerve endings make this distinction?

One clue that enables a nerve ending to respond to a harmful stimulus is that damaging stimuli cause a rupture of cell membranes around living cells. ATP is among the many molecules that rush out of living (but not dead) cells, and is utilized as a pain signal by small diameter nociceptive neurons of the dorsal root ganglia that send axons to the periphery.[22] These neurons possess one of a family of seven proteins called *purinoceptors* (*P2X* receptors) that bind ATP. A specific type of ATP receptor, the *P2X3* protein, responds to pain-producing stimuli by opening an ion channel in the plasma membrane of pain-sensing nerve fibers. In mice with an experimental deletion of this protein, heat or small subcutaneous injections of formalin fail to elicit the pain reactions seen in normal mice.

An ability to feel pain may not immediately seem to us to be a desirable trait, but any serious thought on the matter leads to the conclusion that it is probably one of our most important senses. The importance of pain sensation can be best illustrated by examining the lives of rare people who cannot experience pain. Recently, a small cohort of people insensitive to pain was identified in Pakistan. These individuals placed knives or burning coals on their skin to entertain neighbors. They had numerous bruises and cuts that they were unaware of, and had lesions on the lips and inside the mouth that resulted from accidental bites they had not noticed. Some individuals had broken bones that they had not treated because they caused no pain. Responses to cold, heat, or vibration in these people were normal. All of these symptoms resulted from an inherited abnormality in a portion of a voltage-gated sodium channel protein called *SCN9A* that apparently is important for the function of pain-sensitive axons.[9]

Encapsulated sensory axons

Most of the sensory axons of the skin float freely within the epidermis or in the connective tissue of the dermis or wrap around sensory hairs. Additional axons innervating the skin do not form bare nerve endings, but reside within complicated structures called Meissner's corpuscles or Pacinian corpuscles. The structures possess, in addition to sensory axons, elaborate layers of cells derived from Schwann cells and connective tissue fibroblasts of the perineurium. The precise functions of these encapsulated receptors have proven difficult to determine. Axons within Pacinian corpuscles, for example, respond mainly to rapid vibrations, so it has long been postulated that they might play a role in responding to the roughness of a surface. Conceivably, when you drag your fingertip over sandpaper, this causes your fingertip to vibrate rapidly and perhaps activate Pacinian corpuscles. However, this hypothesis has recently been refuted by electrophysiological recording studies of the skin.[37] Pacinian and Meissner's corpuscles appear to contain pain-sensitive as well as vibration-sensitive axons. A precise identification of the contributions of these structures to sensation must await further study.[31]

3. Sensory Neurons of the Brain

Not all sensory neurons are confined to sensory ganglia. Specialized neurons in the hypothalamus can also be termed sensory neurons. For example, a subset of hypothalamic neurons also possess temperature-sensitive *TRP V1* receptors and monitors the temperature of the blood.[26] Other types of neurons, located in the arcuate nucleus, monitor blood-borne levels of the fat cell hormone, *leptin*. As was noted in Ch. 6, the arcuate nucleus is peculiar because it is adjacent to unusually leaky capillaries and thus is freely exposed to blood-borne molecules, unlike other parts of the brain. Thus, these neurons can monitor blood levels of *leptin*, which informs them about the adipose depots of the body. This allows these neurons to regulate appetite and body weight. These neurons also regulate the autonomic control of the liver and pancreas and exert a powerful influence upon circulating levels of glucose. If the function of these neurons is experimentally restored in genetically obese, diabetic mice, their blood glucose levels rapidly return to normal, even prior to any substantial loss of fat in these mice.[8] It is possible that an abnormal function of these hypothalamic *leptin*-sensitive neurons may contribute to the abnormalities seen in diabetes mellitus.

How does binding of *leptin* to a receptor modulate the function of a neuron? This is accomplished through a signaling pathway called the *JAK-STAT* pathway. *JAK* stands for *Janus Kinase*, i.e., for an enzyme that can phosphorylate other proteins. *STAT* refers to a family of seven proteins noted for an activity called Signaling Activity and Activation of Transcription. Receptors for as many as 35 circulating proteins activate this pathway. These include receptors for *growth hormone, erythropoietin, interleukins, platelet derived growth factor, prolactin*, and of course, *leptin*. Once a receptor binds a ligand, it activates one of four possible associated enzymes called *janus kinases*. These, in turn, facilitate the phosphorylation of a *STAT* protein, which causes each protein to pair up with a partner as a dimer and then to move to the nucleus, where each dimer binds to DNA and modulates transcription.[20] When *leptin* binds to a hypothalamic neuron, its activation can be confirmed because the neuronal nucleus stains positively for the presence of *STAT* (Fig. 8.2).

One of the puzzles concerning the function of *leptin*-sensitive neurons and the control of obesity has involved a relatively rare

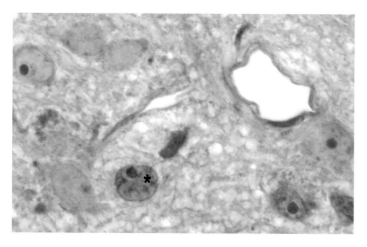

Fig. 8.2. A neuron activated by *leptin* can be identified by the presence of *STAT* in its nucleus (*). Other *leptin*-insensitive neurons, nearby, have unstained nuclei.

inherited disorder called Bardet-Biedl syndrome, which affects about 1 in 100,000 people in the general population. Symptoms of this disorder include obesity, retinal degeneration, hypogonadism, renal dysfunction, and mental retardation. Hypothalamic dysfunction would appear to explain the obesity and hypogonadism in these patients, but what mutational event could also explain the other symptoms?

Recent studies have provided the answers to this puzzle. Mutations in Bardet-Biedl patients all involve proteins that control the function of sensory, non-motile cilia. Since sensory cilia are important for kidney function, this would explain the renal symptoms of these patients. Also, as we shall see, retinal photoreceptor cells possess sensory structures that are derived from cilia, so a ciliary abnormality could also produce retinal problems, as well. But what explains the problems with the hypothalamus?

It turns out that receptors for *leptin* seem to be clustered on the cell membrane of a sensory cilium that is present on hypothalamic neurons. If these receptors are not incorporated into ciliary membranes, or if the cilium does not function, the neuron will no longer respond to *leptin*. This causes obesity and also sterility, since *leptin*-sensitive neurons regulate the synthesis of *gonadotropins* (sexual hormones) by the pituitary gland.[25] Thus, the molecular explanations for this specific type of obesity now seem to be clear. The reasons why the rest of us may become obese are much less certain; no obvious dysfunction of *leptin* or *leptin* receptors accounts for most cases of human obesity.

Reactions of neurons to neurotransmitters

In a larger sense, all neurons of the brain could be termed sensory neurons, since they all respond to *neurotransmitters* released into their environment. Many neurotransmitters affect neuronal function by allowing the entry of ions into a nerve cell. How would this affect the firing rate of a neuron?

All neurons are electrically polarized cells. This means that the outside of the cell membrane is exposed to a higher number of positive ions than the inside. How is this accomplished?

Briefly, neurons possess a membrane-bound *sodium-potassium exchanger* that expels sodium from the interior and imports potassium from the exterior. This in itself would not produce a charge imbalance. What does cause a polarization is that the cell membrane also possesses rather leaky potassium channels and non-leaky sodium channels. This allows potassium to drift out of a cell, down its concentration gradient. A corresponding intake of sodium is not permitted, however, so an excess of positively charged potassium ions builds up on the cell surface.

Binding of many types of neurotransmitters to their receptors can alter the state of electrical polarization because these receptors form parts of *ligand-gated ion channels*. For example, the inhibitory neurotransmitter, *gaba amino butyric acid* (*GABA*), stimulates the uptake of negatively charged chloride ions into the cell, so that the cell interior is even more negatively charged than usual (hyperpolarization). Excitatory neurotransmitters like *glutamate* can bind to several types of receptors; one type is a so-called *ionotropic receptor* that allows sodium and calcium into the cell, thereby decreasing the charge imbalance between the inside and outside of the cell (depolarization). The total sum of all the inhibitory and excitatory inputs onto the cell determines whether the cell is relatively hyperpolarized or depolarized.

Why would a depolarization cause an electrical impulse to appear in the axon of a neuron? This is because an axon (but not a dendrite or a nerve cell body) possesses specialized *voltage-sensitive sodium channels* that open when the electrical potential across the cell membrane falls to a low level. This causes massive amounts of sodium to rush into an axon, which triggers voltage-sensitive channels farther along the axon to open, which causes more sodium to enter, and on this process goes until the end of the axon is reached. The wave of depolarization (action potential) that flashes along the length of an axon is soon corrected by the re-adjustment of all of these channels, which return the cell to its previous state, ready to respond again.

This process of propagating an action potential differs quite substantially between myelinated and non-myelinated axons. Myelinated

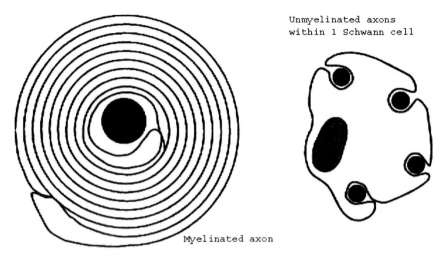

Unmyelinated axons
within 1 Schwann cell

Myelinated axon

Fig. 8.3.   Diagram comparing a myelinated axon with unmyelinated axons.

axons are surrounded by numerous layers of cell membrane, derived from oligodendrocytes (CNS) or Schwann cells (PNS) (Fig. 8.3). Unmyelinated axons still occupy protected niches within the cytoplasm of glial cells, but are not insulated by the elaborate layers of myelin. When an action potential travels along the membrane of an unmyelinated axon, it must slowly and sequentially activate a series of voltage-dependent sodium channels one by one. This is not true for a myelinated axon.

In a myelinated axon, voltage-sensitive sodium channels are not distributed uniformly along the axon. Membranes located just beneath the myelin sheath have relatively few sodium channels (about 25 per square micron of membrane) (Fig. 8.4). However, membranes located at the intervals between one myelinating glial cell and another (this interval is called a node of Ranvier) possess far more sodium channels (1000 per square micron of membrane). As a result, flows of ions produced by an axon potential are much greater at a node of Ranvier than at other membrane sites. The electrical disturbance at a node of Ranvier is mainly propagated through the watery extracellular fluid than along the membrane. This jumping mode of conduction, called saltatory conduction after the Latin word for

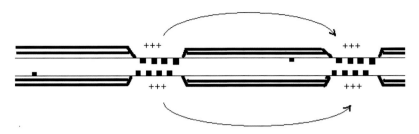

Fig. 8.4. Diagram of a myelinated axon, showing how voltage-dependent sodium channels (small squares) are preferentially located at the nodes of Ranvier where the myelin sheath is interrupted by a small gap between glial cells.

jump, is much faster than the rates of conduction in unmyelinated axons.[33]

Not all receptors for neurotransmitters alter the function of ion channels. What is more, a single type of neurotransmitter can affect multiple types of receptors that have differing effects upon neuronal function. *Glutamate* receptors, for example, can be divided into either *ionotropic receptors* that alter the flow of ions into cells or into *metabotropic receptors* that are associated with *G proteins* and which alter levels of second messengers in cells. Receptors for *acetylcholine*, another widespread neurotransmitter that functions both in the brain and at neuro-muscular junctions, follow a similar pattern. Two types of so-called *nicotinic receptors* for *acetylcholine* provoke the entry of sodium into cells (and are also sensitive to the plant molecule, *nicotine*). Another five types of so-called *muscarinic receptors* for *acetylcholine* are associated with *G proteins* that alter levels of second messengers like *cAMP* or *inositol triphosphate*.[17] This multiplicity of both neurotransmitters and receptors makes the study of neural tissue very complicated.

To make things even worse, yet another level of complexity has become known since the late 1980's. At about this time, techniques for staining synapses and receptors were fully developed. When these techniques were applied to brain tissue, however, a disturbing paradox became apparent. Frequently, large numbers of receptors for a neurotransmitter like *dopamine* or *norepinephrine* could be detected in a given brain region; however, the number of synapses containing these transmitters in the same brain could be much smaller! This apparent "mis-match" between synapses and receptors was hard to understand. Even more puzzling, electron microscopy revealed that receptors for *peptide neurotransmitters* like *endorphin* or *enkephalin* are almost always located a considerable distance *away* from the post-synaptic portions of neuronal membranes!

The accumulation of data like these has lead to a hypothesis that much of information transfer within the nervous system may be communicated *extra-synaptically*; in other words, neurons can simply secrete neurotransmitter in all directions into extracellular fluid, which then carries molecules to receptors that can be located at both synaptic and non-synaptic parts of a neuron. This type of information processing, termed volume transmission, is hypothesized to be less rapid and less precise than synaptic transmission, but is nevertheless important for understanding the brain.[12]

Besides binding neurotransmitters, neurons also possess receptors for a variety of growth factors like nerve growth factor and brain-derived growth factor that promote their survival. The receptors that bind these growth factors are called *receptor TyRosine Kinases* (*Trk's*) because, when each receptor binds a ligand, it acquires the ability to add phosphate groups to tyrosine amino acids.

A complex series of events occurs when these receptors bind a ligand. Two receptor molecules become associated with each other in the plane of the plasma membrane, add phosphate groups to each other, and form a dimer. Each dimer then can activate a series of other proteins (e.g., a kinase called *Raf* and a *GTP*-binding protein called *Ras*) that can trigger the movement of transcription regulating proteins into the nucleus or alter the activity of anti-apoptotic proteins like *Bcl2*.[16] The *Ras* protein, which activates a complex pathway called the *Mitogen Activated Protein Kinase* (*MAPK*) pathway, has a particularly important role in controlling cell proliferation. It was first discovered in a Rat Sarcoma tumor (hence the name *Ras*); mutations in *Ras* that increase its activity are found in 20–30% of all human cancers.[4]

4. Sensory Functions of the Inner Ear

Sensory cells of the inner ear are contained within discrete portions of a tortuous, membranous tube filled with a fluid called endolymph. All of the sensory cells are designed to detect vibrations in endolymph, for differing purposes and sensations.

A glance at a high school biology book will remind you of the divisions of the inner ear. The central part of the inner ear is composed of two irregular, conjoined sacs called the utricle and the saccule. Sprouting from the saccule at one end of the inner ear is a long, coiled tube (the cochlear duct) contained within a layer of bone that rather resembles the shell of a snail. This portion, of course, is the cochlea, which detects sounds. At the other end of the inner ear, sprouting from the utricle, are three hoop-shaped tubules called semicircular canals. Sensory cells at the bases of these tubules detect angular movements of the head. Finally, within the utricle and saccule themselves, other sensory cells are assembled into small patches called maculae. These cells monitor the position of the head relative to the surrounding gravity field. How do these cells accomplish these tasks?

All of these sensory cells have many features in common. The simplest of the sensory cells are found in the maculae.

Maculae

Each macula contains several thousand sensory cells, called hair cells, covered by a gelatinous protein matrix (Fig. 8.5). Embedded within this gelatinous matrix are dozens of small, rock-shaped particles called otoconia.

The otoconia function as weights that press down upon the underlying protein matrix and upon the sensory cells. If the position of the head becomes tilted away from vertical, gravity will shift the positions of the otoconia and cause the underlying sensory cells to react.

How do the sensory cells keep track of the positions of the otoconia? Each sensory cell possesses 20–50 very long microvilli that

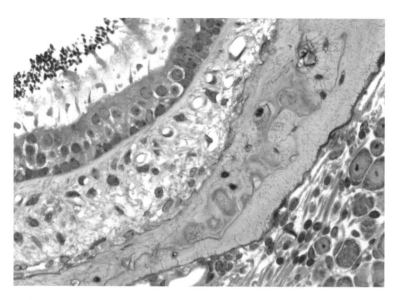

Fig. 8.5. View of a macula, showing (upper left) dark-staining otoconia, an under-lying protein-rich matrix, and sensory cells with pink-stained tufts of stereocilia extending into the matrix. Cells that do not bear sensory stereocilia are called accessory cells. Beneath the sensory epithelium is a thin basement membrane, loose connective tissue rich in capillaries, and bone. At the lower right of the figure, neuronal cells of the vestibular ganglion, possessing large, dark-stained nucleoli, are visible. These neurons innervate the macula.

are commonly called stereocilia by auditory researchers (each cell also possesses a single, non-motile cilium of uncertain function). Like most microvilli, these stereocilia contain an axis of *actin* protein formed into a stiff bundle by another protein called *fimbrin*. When movement of the otoconia causes the stereocilia to bend, the sensory cells respond by releasing neurotransmitters (*glutamate*) from their basal surfaces. This causes the activation of adjacent sensory axons that originate from nearby nerve cells of the vestibular ganglion.

How does a cell "know" that its stereocilia have been bent? This question has only been resolved relatively recently. Each stereocilium possesses a filament composed of *cadherin-23* at its tip. These filaments are called tip links; each tip link interconnects adjacent stereocilia (Fig. 8.6).[18]

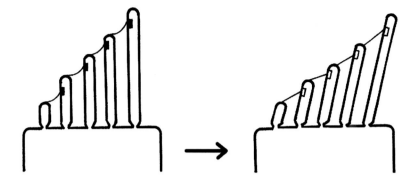

Fig. 8.6. Diagram of the stereocilia of a sensory hair cell of a macula. When stereocilia are vertically positioned, connecting tip links are slack and the ion channels they connect to (black squares) are closed. When the tip links tighten, their associated ion channels (white squares) are opened.

When the stereocilia are positioned vertically, the tip link strands are loosened; when stereocilia become bent, the tension on the tip links increases. The tip links, in turn, tug on the membrane and cause the opening of membrane-associated potassium channels that change the electrical activity of the sensory cell. The identity of these mechanically activated channels is still being debated. There is some good initial evidence that they are formed from a specific type of *TRP* receptor called *TRPML3*.[10]

The functionality of all of these components is well illustrated in an experimental strain of mice deficient for a protein called *otoconin-90*. This protein forms the core of otoconia. It is apparently secreted from cells adjacent to the macula, and binds to proteins like *otogelin* and *otoancorin* that make up the gelatinous matrix of a macula. Once incorporated into this matrix, *otoconin-90* facilitates the precipitation of crystals of calcium carbonate that compose the non-organic portion of otoconia. If *otoconin-90* is absent, otoconia either fail to form or are badly misshapen. Mice deficient in *otoconin-90* appear to hear normally (cochlear function is intact), but cannot balance very well when held in the hand and hold their heads at a peculiar angle.[38] Such mice lack a good sensation of gravity.

Cristae

Other patches of sensory epithelia, found at the base of each semicircular canal, are called cristae. Instead of being flat like maculae, these cristae form ridges of epithelial cells that project up into the lumen of each semicircular canal. However, these cells are also covered by a gelatinous matrix, called a cupula, which seems to have a function somewhat similar to that of the proteins covering a macula (Fig. 8.7). The gelatinous cupula acts rather like the swinging door of a kitchen restaurant that opens and closes to allow the passage of busy waiters.

When a semicircular canal is moved around its axis by a head movement, endolymphatic fluid rushes around inside the canal and pushes on the cupula. This also exerts tension upon the stereocilia of

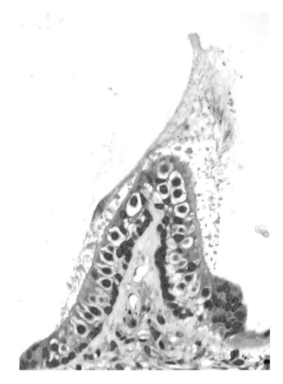

Fig. 8.7.   View of a crista ampullaris at the bottom (ampulla) of a semicircular canal.

sensory cells and signals the brain that there has been a rotational movement of the head. This signaling takes place more often than we might recognize: each time we turn our head to track an object moving across our field of vision, the semicircular canals measure our head rotation and initiate signals that coordinate the movement of our eyes with the movement of our head so that the visual object remains in view. This is easily illustrated by simply holding your finger in front of you, focusing your gaze upon it, and noticing what happens to your eyes when you turn your head from side to side. Your eyes will "automatically" track your finger and focus on it in spite of your head movement.

Organ of Corti

A stripe of specialized sensory cells can also be found running along a portion of the inner surface of the spiraling cochlear duct. In cross section, this collection of cells, called the organ of Corti, has a distinctive appearance because it is covered over by a gelatinous flap of non-cellular material called the tectorial membrane (Fig. 8.8).

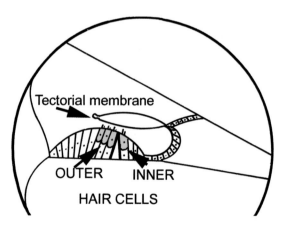

Fig. 8.8. Diagram of the Organ of Corti, showing the centrally located cochlear duct with its specialized, sensory hair cells and the tectorial membrane. The wedge-shaped spaces above and below the cochlear duct are filled with perilymph rather than endolymph. The stria vascularis is located just to the left of the organ of Corti.

The tectorial membrane contains some of the molecules that are also found in the extracellular matrix that surrounds cartilage cells: *proteoglycans* and *type II collagen* (plus *types V* and *IX*).[24] These molecules presumably contribute to the flexibility of the tectorial membrane, which bounces up and down within the endolymph in response to vibrations induced by sound waves.

When the tectorial membrane comes in contact with the stereocilia of the sensory cells, called inner hair cells, this mechanical stress causes the same reactions as seen in the other hair cells of the ear. Sound waves reach the fluid of the cochlear duct when a middle ear bone, the stapes, presses upon a structure called the oval window. This causes vibrations to form in the perilymph, a fluid found in the bony cavity that surrounds the cochlear duct. When vibrations travel across the epithelium comprising the cochlear duct, they disturb the endolymph and cause movements of the tectorial membrane and of the basilar membrane beneath the hair cells (Fig. 8.9).

In addition to *collagens*, the tectorial membrane also contains proteins unique to the inner ear like *α-tectorin* and *otogelin*. Mutations in these proteins can cause hearing impairment, and may be important clinically, since over half of all childhood hearing disabilities are due to inherited mutations in one or more inner ear proteins.[19] Other examples of deafness-producing mutations involve the *TRPML3* mechanoreceptor protein and a specialized type of *myosin, myosin VIIa*, which binds to the *actin* in stereocilia.[10,24]

A more widespread cause of deafness is not the inheritance of an abnormal protein, but the gradual damage and cell death that occur among the hair cells of the cochlea. Hearing impairment due to these types of events affects 250 million people worldwide and is the second most common disability. Curiously, sensory cells of the inner ears of birds also undergo cell loss, but these losses can be compensated for by the transformation and division of adjacent supporting cells into new sensory cells. If the reasons why mammals, but not birds, fail to regenerate hair cells could be understood, it may someday be possible to avoid the age-related hearing loss that occurs so commonly.

A detailed examination of the physiology of hearing would be a very complex topic and is beyond the scope of this book. For

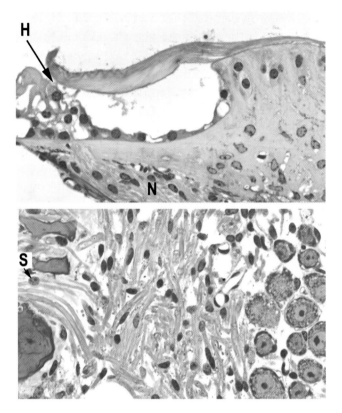

Fig. 8.9. TOP: View of the organ of Corti, showing a sensory inner hair cell (H) in contact with the tectorial membrane. Sensory nerve fibers (N) course through the basilar membrane to innervate the hair cells. BOTTOM: Magnified view of the spiral ganglion, located below and to one side of the organ of Corti. Neurons (far right) must send sensory axons through pores in a bony shelf that supports the organ of Corti. The nucleus of a Schwann cell (S) can be seen in association with sensory fibers.

example, the outer hair cells respond to vibrations of the tectorial membrane not with a sensory response, but with energetic expansions and contractions that amplify vibrations and make the inner hair cells more sensitive to sound. Also, specializations of the basilar membrane are responsible for why some hair cells are responsive to low frequency sounds and others to higher frequency sounds.

One question, however, is of interest to us: what is the source of the endolymph that bathes all of the sensory cells of the inner ear? As it turns out, endolymph is produced by a patch of very peculiar epithelial cells that is found to one side of the organ of Corti within the cochlear duct. This patch of epithelium is called the stria vascularis (Fig. 8.10). It is very strange, because it is the only epithelium within the body that contains capillaries! Why does this occur?

It makes functional sense for this epithelium to possess capillaries. These capillaries allow for the extraction of lots of fluid and also for the transport of potassium, which is present in high concentrations within the endolymph. However, most epithelia are walled off from capillaries of the underlying connective tissue by a basal lamina that prevents movement of cells across it. How does this principle come to be violated in the stria vascularis?

Recent work suggests that, surprisingly, the peculiar anatomy of the stria vascularis is the creation of melanocytes that migrate into the epithelium during development. Melanocytes arise from a specific portion of the neural tube called the neural crest; as they mature, they

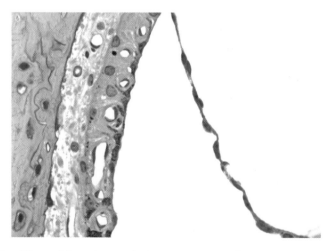

Fig. 8.10.   View of the stria vascularis. Numerous capillaries can be seen within the epithelium (these are empty-appearing because blood cells have been flushed away with fixative). Several cells within the epithelium that contain dark melanin granules can be seen at the bottom of this picture. Bone is also visible at the far left.

migrate into epithelia like the skin to provide a protective pigment (melanin) that shields skin cells from damaging effects of ultraviolet light. In the specific case of the stria vascularis, these pigmented melanocytes somehow stimulate the development of a capillary network within the epithelium. If melanocytes are experimentally damaged during development, the stria vascularis degenerates and the production of endolymph is impaired.[14] How melanocytes accomplish this task is still not well understood.

5. Sensory Functions of the Eye

The vast majority of the light-sensitive cells of the retina of the eye are specialized neurons called rod cells. They have a distinctive anatomy, with the cytoplasm divided into three regions: one region contains the nucleus, another region (the inner segment) contains most cytoplasmic organelles, and a third region (the outer segment) contains about one thousand flattened, hollow, membranous discs (Fig. 8.11).

The junction between the inner segment and outer segment of a cell is marked off by the presence of a microtubule-containing basal body like that seen beneath most cilia of the body. This signifies that the rod outer segment is nothing more than a highly modified cilium. How does it function to react to light?

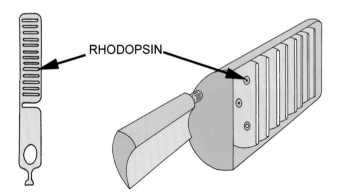

RHODOPSIN

Fig. 8.11.   Diagrams of a rod cell, showing the membranous discs of the outer segment and the locations of the visual pigment, *rhodopsin.*

In the retina, light penetrates through a number of layers of nerve cells and then passes longitudinally through the axis of a rod cell (Fig. 8.12). When it traverses the outer segment of a rod, the presence of so many membranous discs makes it likely that a photon of light will interact with a molecule in at least one disc. The molecule that reacts to light is called *rhodopsin*, and is composed of a protein (*opsin*) and form of vitamin A called *retinal*. *Retinal* contains many double bonds between its 19 carbon atoms, which increase its ability to absorb light, just like many other colored compounds such as dyes or *chlorophyll*. When *retinal* absorbs a photon of light, this causes the molecule to change its shape; as a result, *retinal* dissociates from *opsin*.

Opsin that has been freed of its association from *retinal* has the ability to function like a *G-protein linked receptor*. It interacts with a *G-protein* called *transducin*. *Transducin* activates an enzyme called *cGMP phosphodiesterase*, which decreases intracellular concentrations of *cyclic GMP* (*guanosine monophosphate*). Finally, as a last step in the reaction to light, *cGMP*-sensitive sodium channels in the plasma membrane of the cell close, preventing the uptake of sodium. This causes the cell to become hyperpolarized, so that it generates fewer action potentials than normal in the presence of light. It is curious that what we perceive as the activation of the retina by light is actually an inhibition of rod cells. Rod cells actually fire more rapidly in the dark than in the light.

What causes rod cells to acquire their peculiar anatomy? This appears to be directly related to the synthesis of *rhodopsin* itself. *Rhodopsin* is first synthesized in the inner segment and then is directed into early endosomes near the ciliary basal body by a protein called *SARA* (*Smad Anchor for Receptor Activation*). *SARA* apparently is very important in most cells for the function of early endosomes. In rod cells, *rhodopsin* and *SARA* are passed from endosomes to membranous discs via an interaction with the *SNARE* protein *syntaxin 3*. *Rhodopsin* is highly enriched in disc membranes, constituting 90% of membrane protein. If the movement of *rhodopsin* from early endosomes is disrupted by altering the function of *SARA*, gross disturbances in the shape and numbers of membranous discs

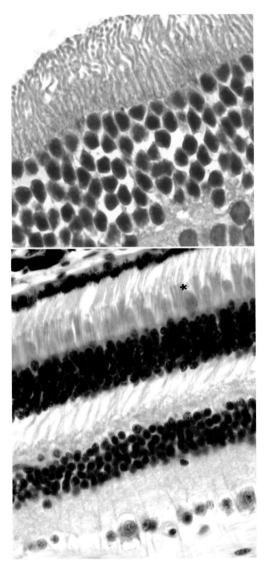

Fig. 8.12. TOP. View of a mouse retina, showing the tube-like outer segments of rods at the top of the picture. Note how the nuclei of the rod cells stain very darkly in comparison to the nuclei of adjacent neurons (lower right). BOTTOM. Low magnification view of a human retina, showing the more truncated appearance of the outer segment of a cone cell (*) nestled between rod cells. The nuclei at the bottom of the picture belong to ganglion cells and the nuclei forming a layer just above them belong to information-processing neurons like bipolar cells.

ensue.[7] So, the strange morphology of the sensory cilium of rod cells seems to be a direct consequence of the synthesis of *rhodopsin*. The membranous discs of rods move up to the tips of each outer segment within ten days; they are subsequently shed to be phagocytized by nearby pigment epithelial cells. Thus, for uncertain reasons, these light-sensing organelles have an unusually rapid turnover in rods.

Rod cells of nocturnal animals like mice and other rodents have another specialization. You may recall from Ch. 1 that highly active genes of euchromatin are typically found in the center of the nucleus and that inactive, condensed heterochromatin is typically located close to the nuclear envelope. In rod cells of mice, this nearly universal pattern of active genes is inverted, so that condensed chromatin is found at the center rather than at the periphery of each nucleus. Computer modeling suggests that the denser cores of the resultant nuclei function as tiny lenses that direct photons of light towards the membranous discs of the outer segments.[27] This is yet another example of the marvelous adaptations created by mother nature to improve the functions of cells.

The 125 million rod cells of each retina are exquisitely sensitive to dim light, but do not provide for perception of different colors. This task is accomplished by cone cells, which are distinguishable from rods by their shortened, cone-shaped outer segments. Cone cells of humans express genes for 3 different types of *opsins*; variations in the amino acid sequences between these types impose slightly different electrical charges upon the associated retinal molecule and "tunes" it to respond maximally to either blue, green, or red light.[21] How ancestors of human beings evolved to retain full three-color vision, while most other mammals lost this full capacity during evolution from reptiles, is still a matter of some dispute.

Rods and cones are not the only cells in the retina that directly respond to light. A subset of ganglion cells located close to the inner surface of the retina also possess their own form of *opsin* (*melanopsin*) and react to light independently of any input from rods or cones. These cells apparently mediate the less sophisticated, simple determination of whether it is daytime or nighttime, and communicate this information to the suprachiasmatic nucleus of the hypothalamus, a collection of neurons that imposes day-night differences (circadian rhythms) upon many physiological events.[3]

Finally, sensory cells in the eye may also play a part in responses to *non-visual* stimuli. Abundant evidence shows that many species of birds and mammals have the ability to locate their positions in space with reference to the earth's magnetic field. Herds of grazing cattle and deer, for example, can be reliably observed to position themselves parallel to magnetic lines of force.[2] This ability is particularly important in mole rats, which burrow underground and have no visual cues to orient themselves. Crystals of magnetite (iron oxide) can be found in the corneas (surface epithelium covering the pupil) of these animals, and corneal anesthesia renders them insensitive to magnetic fields.[34] Thus, the cornea may be the location of heretofore unappreciated sensory cells that respond to magnetism.

6. Olfaction

Sensory responses to air-borne odorants originate in olfactory neurons that are located within a pseudostratified columnar epithelium covering the upper surface of the nasal cavity (Fig. 8.13).

The actual sensory cells within the olfactory epithelium make up a relatively small portion of the cells present; they possess non-motile sensory cilia at their apical portions that bind odorants. Curiously, these nerve cells are among the few in the body that are regularly replaced after injury or death; basally located stem cells can differentiate into neurons and replace them.[32] The surrounding accessory cells synthesize growth factors like *IGF1* that may regulate the renewal and growth of olfactory neurons.[30] The surface of the epithelium is continually bathed by secretions from the nearby Bowman's glands.

How do olfactory neurons respond to odors? The functional capacity of these cells to detect odors varies greatly from species to species, but the basic mechanisms are fundamentally the same. In mice, a family of over 1000 individual proteins has been detected that constitute the olfactory receptor family. Each protein is a *G-protein coupled receptor*; when the receptor binds an odorant, this leads to the activation of *adenyl cyclase* and an increase in cytoplasmic *cAMP*

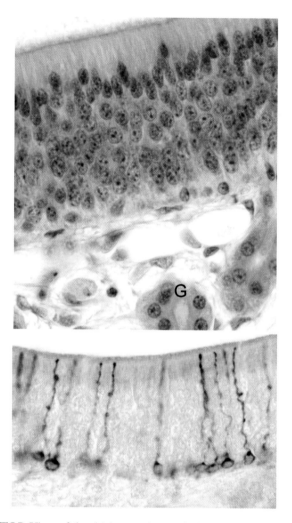

Fig. 8.13.   TOP. View of the thick, pseudostratified columnar epithelium found in the olfactory mucosa of a human. Most of the cells found here are either stem cells or accessory cells. Beneath the epithelium, Bowman's glands (G), formed from simple cuboidal epithelial cells, can be seen. BOTTOM. Mouse olfactory mucosa, stained for *Olfactory Marker Protein* to illustrate the bipolar olfactory neurons. Courtesy of Dr. Eric Walters, Dept. Biochemistry, Howard University.

(see Fig. 8.1). This response, in turn, causes the opening of sodium channels in the plasma membrane and a depolarization of the cell.[5] Humans possess a much more limited variety of olfactory receptor proteins (about 350) and are less able to distinguish between various odors than animals like mice that depend heavily upon olfactory cues about their environment.

Each one of the millions of olfactory neurons in the nose binds only one type of odorant molecule. An activated neuron sends a signal to the brain via an axon that terminates in only 1 out of about 2000 spherical structures in the olfactory bulb that are called olfactory glomeruli (Fig. 8.14). Within each glomerulus, axons synapse upon processes from additional neurons called mitral cells that transmit their information into the olfactory cortex. A single cortical neuron can receive information from as many as 50 olfactory glomeruli, so a great degree of convergence of information takes place. Presumably, a complex smell (burning wood, steaming soup), containing many different odorant molecules, activates numerous glomeruli that send signals to many different cortical neurons, leading to a very complicated interpretation of what is in the air.

The olfactory epithelia and bulbs are not the only olfactory organs of mammals. In many mammals, another organ called the vomeronasal organ plays a major role in reacting to specific odorants called pheromones that are secreted into the environment by animals. These odorous components of sweat or urine have a powerful influence on behavior in many mammals. Sensory cells of the vomeronasal organ are located in small pits associated with the nasal septum. In mice, each cell expresses one out of a several families of about 300 proteins that bind pheromones.[23] The vomeronasal cells project to the accessory olfactory bulb and other structures quite different from those involved in detecting non-pheromonal odors.

A range of behaviors are affected by pheromones and the vomeronasal system. Male mice will frequently attack a strange male placed into their cages; if the strange male is castrated, however, it will not be attacked. The involvement of testosterone-stimulated pheromones in eliciting attacks can be demonstrated by simply "painting" the back of a castrated male with the urine of an intact

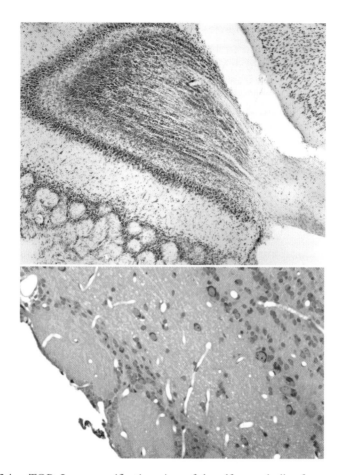

Fig. 8.14.   TOP: Low magnification view of the olfactory bulb of a mouse brain. The anterior part of the bulb is to the left, and the posterior portion, which connects to the remainder of the brain, is to the right. The small, pale-staining circles at the bottom of the bulb are olfactory glomeruli. BOTTOM: Higher magnification view of olfactory glomeruli (left) and layers of neurons that receive inputs from the glomeruli (right).

male, which will restore attacks upon the urine-painted male. Two urinary proteins have been identified that serve as the pheromones promoting male-male aggression.[6]

Another striking effect of pheromones can be seen when a female mouse is exposed to the urine of a male mouse unfamiliar to her.

This circumstance prevents pregnancy and is called the "Bruce effect" after Hilda Bruce, who first discovered it. The Bruce effect is mediated by neurons of the vomeronasal organ that activate a hypothalamic circuit that suppresses the release of *prolactin*, which in mice is necessary for pregnancy.[35] The ability of some animals to recognize the odor of specific individuals thus seems dependent upon the vomeronasal organ.

Do humans utilize this organ? A small vomeronasal organ does exist in the human nasal mucosa, but analysis of the human genome shows that only 4 of 200 pheromone receptor genes have survived intact during evolution. The remaining genes for human pheromone receptors have acquired DNA alterations that have rendered them into non-functional "pseudogenes." Thus, it is likely that the vomeronasal organ is vestigial in humans.[13]

7. Taste

The sensation of taste is mediated by collections of sensory cells (taste buds) found in the stratified squamous epithelium covering the tongue (Fig. 8.15). Each taste cell is surrounded by basal stem cells and accessory cells in each taste bud. When a molecule dissolved in saliva binds to receptor proteins found upon the apical microvilli of each taste cell, a signal to the brain is initiated. Like odorant receptors, taste receptor proteins are linked to a *G-protein* (called *gustducin*) that increases cellular levels of GTP. GTP-sensitive ion channels thus depolarize the cell when the cell binds a tastant.

Humans can taste 5 basic taste modalities: sweet , bitter, sour, salty, and an additional savory flavor that takes its name from the Japanese word *umami*. These molecules bind to specific receptor proteins present on the apical cell membranes of taste cells.

Sweet tastes are elicited by molecules like sugars and alcohols. These substances bind to cell membrane proteins called *T1R*'s. Three types of these sweet-sensitive receptors exist; they are found in about 30% of all taste bud cells.

Bitter tastes are elicited by a wide range of substances like *strychnine* or *cyanide*. Many of these are produced by plants and are

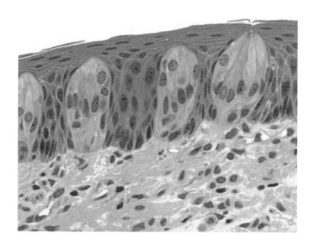

Fig. 8.15. A view of 4 taste buds embedded within the stratified squamous epithelium of the tongue. The surrounding conventional epithelial cells stain pinker because they contain higher amounts of the intermediate filament, *keratin*. Taste cells contact saliva at an apical opening in the epithelium called the taste pore (visible at the top of the taste bud at the far right).

poisonous. These substances bind to *T2R proteins*. Twenty-four types of these proteins have been found in humans; they can identify many toxins that may be present in food.[28]

Sour tastes are related to high concentrations of acid (hydrogen ions) in a food. These ions are detected by a newly discovered protein called *PKD2L1*. This protein is one of a family of proteins that are also found in sensory cilia in kidney cells.[15] Salty tastes seem to be detected by some type of protein that forms a sodium channel in the cell membrane. Several candidate proteins have been suggested to act as sensors for savory tastes. They include proteins called the *T1R3* protein and the *metabotropic glutamate receptor 4*.[28]

One curious feature of taste buds is that their development and maintenance is totally dependent upon signals from sensory axons that innervate them. If the nerves to a rat's tongue are cut, the taste bud cells undergo apoptosis, degenerate, and disappear. Upon regrowth and recovery of the nerves, taste buds reappear once more.

Some unknown molecule, released from sensory axons, must have a trophic influence that supports the differentiation and stability of taste cells. This neural signal, whatever it is, stimulates the taste cells to produce a transcription regulating factor called *Sox2*, which is needed for the development of taste cells.[29]

References

1. Bautista DM, *et al.* (2006) TRPA1 mediates the inflammatory actions of environmental irritants and proalgesic agents. *Cell* 124: 1269–1282.

2. Begall S, *et al.* (2008) Magnetic alignment in grazing and resting cattle and deer. *Proc Natl Acad Sci USA* 105: 13451–13455.

3. Berson DM, Dunn FA, Takao M (2002) Phototransduction by retinal ganglion cells that set the circadian clock. *Science* 295: 1070–1073.

4. Bos J (1989) Ras oncogenes in human cancer: a review. *Cancer Res* 49: 4682–4689.

5. Buck LB (2004) Olfactory receptors and odor coding in mammals. *Nutr Rev* 62: S184–S188.

6. Chamero P, *et al.* (2007) Identification of protein pheromones that promote aggressive behavior. *Nature* 450: 899–905.

7. Chuang J-Z, Zhao Y, Sung C-H (2007) SARA-regulated vesicular targeting underlies formation of the light-sensing organelle in mammalian rods. *Cell* 130: 535–547.

8. Coppari R, *et al.* (2005) The hypothalamic arcuate nucleus: a key site for mediating leptin's effects on glucose homeostasis and locomotor activity. *Cell Metab* 1: 63–72.

9. Cox JJ, *et al.* (2006) An SCN9A channelopathy causes congenital inability to experience pain. *Nature* 444: 894–898.

10. Cuajungco MP, Grimm C, Heller S (2007) TRP channels as candidates for hearing and balance abnormalities in vertebrates. *Biochim Biophys Acta* 1772: 1022–1027.

11. Folgering JH, *et al.* (2008) Molecular basis of the mammalian pressure-sensitive ion channels: focus on vascular mechanotransduction. *Prog Biophys Mol Biol* 97: 180–195.

12. Fuxe K, *et al.* (2007) From the Golgi-Cajal mapping to the transmitter-based characterization of the neuronal networks leading to two modes

of brain communication: wiring and volume transmission. *Brain Res Rev* 55: 17–54.

13. Grus WE, *et al.* (2005) Dramatic variation of the vomeronasal pheromone receptor gene repertoire among five orders of placental and marsupial mammals. *Proc Nat Acad Sci USA* 102: 5767–5772.

14. Hoshino T, *et al.* (2000) Cochlear findings in the white spotting (Ws) rat. *Hearing Res* 140: 145–156.

15. Huang AL, *et al.* (2006) The cells and logic for mammalian sour taste detection. *Nature* 442: 934–938.

16. Kaplan DR, Miller FD (2000) Neurotrophin signal transduction in the nervous system. *Curr Opin Neurobiol* 10: 381–391.

17. Kester M, *et al.* (2007) Elsevier's Integrated Pharmacology. Elsevier: Philadelphia, p. 87.

18. Kierszenbaum AL (2007) Histology and Cell Biology. An Introduction to Pathology. Philadelphia: Mosby, p. 286.

19. Legan PK, *et al.* (2005) A deafness mutation isolates a second role for the tectorial membrane in hearing. *Nature Neuroscience* 8: 1035–1042.

20. Mertens C, Darnell Jr JE (2007) SnapShot: JAK-STAT signaling. *Cell* 131: 612.

21. Nei M, Zhang J, Yokoyama S (1997) Color vision of ancestral organisms of higher primates. *Mol Biol Evol* 14: 611–618.

22. North RA (2002) Molecular physiology of P2X receptors. *Physiol Rev* 82: 1013–1067.

23. Pantages E, Dulac C (2000) A novel family of candidate pheromone receptors in mammals. *Neuron* 28: 835–845.

24. Raphael Y, Altschuler RA (2003) Structure and innervation of the cochlea. *Brain Res Bull* 60: 397–422.

25. Seo S, *et al.* (2009) Requirement of Bardet-Biedl syndrome proteins for leptin receptor signaling. *Human Molec Genet* 18: 1323–1331.

26. Sharif-Naeini R, Ciura S, Bourque CW (2008) TRPV1 gene required for thermosensory transduction and anticipatory secretion from vasopressin neurons during hyperthermia. *Neuron* 58: 179–185.

27. Solovei I, *et al.* (2009) Nuclear architecture of rod photoreceptor cells adapts to vision in mammalian evolution. *Cell* 137: 356–368.

28. Sugimoto K, Ninomiya Y (2005) Introductory remarks on umami research: candidate receptors and signal transduction mechanisms on umami. *Chemical Senses* 30: i21–i27.

29. Suzuki Y (2008) Expression of Sox2 in mouse taste buds and its relation to innervations. *Cell Tiss Res* 332: 393–401.

30. Suzuki Y, Takeda M (2002) Expression of insulin-like growth factor family in the rat olfactory epithelium. *Anat Embryol* 205: 401–405.

31. Vega JA, *et al.* (2009) The Meissner and Pacinian sensory corpuscles revisited: new data from the last decade. *Microsc Res Tech* 72: 299–309.

32. Walters E, *et al.* (1996) *LacZ* and OMP are co-expressed during ontogeny and regeneration in olfactory receptor neurons of OMP promoter-*lacZ* transgenic mice. *Int J Devel Neuroscience* 14: 813–822.

33. Waxman SG, Ritchie JM (1993) Molecular dissection of the myelinated axon. *Ann Neurol* 33: 121–136.

34. Wegner RE, Begall S, Burda H (2006) Magnetic compass in the cornea: local anaesthesia impairs orientation in a mammal. *J Exp Biol* 209: 4747–4750.

35. Wersinger SR, *et al.* (2008) Inactivation of the oxytocin and the vasopressin (Avp)1b receptor genes, but not the Avp1a gene, differentially impairs the Bruce effect in laboratory mice. *Endocrinology* 149: 116–121.

36. Wetzel C, *et al.* (2007) A stomatin-domain protein essential for touch sensation in the mouse. *Nature* 445: 206–211.

37. Yoskioka T, *et al.* (2001) Neural coding mechanisms underlying perceived roughness of finely textured surfaces. *J Neurosci* 21: 6905–6916.

38. Zhao X, *et al.* (2008) Otoconin-90 deletion leads to imbalance but normal hearing: a comparison with other otoconia mutants. *Neuroscience* 153: 289–299.

EPILOGUE

What general lessons have we learned from all of the preceding information? One obvious theme is that biology in general, and cell biology in particular, is being transformed from a descriptive into a mechanistic science. In previous decades, cell biologists could create descriptive names for events (e.g., prophase of mitosis), but the mechanisms for these events were mysteries. Now we are beginning to understand the mechanisms that create these biological events (e.g., the onset of DNA synthesis is initiated by the phosphorylation of *Rb protein* by *cyclinD* and *cyclin-dependent kinase*). This knowledge of mechanisms not only leads to a greater understanding of biological events, but gives us tools to precisely manipulate them.

An ability to manipulate cells has many implications for both clinical biology (medicine) and for theoretical biology. The challenges to human health that confront physicians have changed greatly since the beginning of the 20th century. Infectious diseases used to pose the major threats to health; now, microbes have become less of a hazard due to the many triumphs of microbiologists, hygienists, and developers of antibiotics.[4] In the 21st century, major health problems stem not from infections but from the gradual onset of cell dysfunctions that lead to disorders like heart disease, diabetes, neurological disorders, cancer, and arthritis. These disorders need to be addressed, not with antibiotics, but with approaches that correct cell function. For example, if pluripotent stem cells could be created from the connective tissue of an individual with diabetes, and if those stem cells could be coaxed into differentiating into pancreatic beta cells that secrete insulin, a potential cure for diabetes could be close to realization.[3] Another example involves the growing understanding of the cell biology of dopaminergic neurons. It has recently been shown that the rhythmic firing of these neurons in the substantia nigra of the brain requires the repeated influx of calcium through a specific type of calcium channel in the plasma membrane. This calcium influx, while necessary for the function of the cell, also gradually damages the cell and may be a major cause of the dopamine deficiency that occurs in Parkinson's disease. Blockade of these channels with drugs called dihydropyridine antagonists may prove to be a method for slowing the progression of cell damage in Parkinson's disease.[1] Thus, abstract curiosity about how cells differentiate or how neurons create their own unique firing patterns may have unexpected, practical benefits for medicine. Some of the facts presented in this book may also prove to have practical applications in the future.

The fruits of the study of cells may also be seen in less practical branches of biology. The evolution of the vast array of structures in the animal kingdom — extremes of form found in anteater's noses, elephant's trunks, giraffe's necks, etc. — has always been presumed to originate from minor variations in the DNA and morphology of an offspring that could be selected for and magnified by environmental pressures. But where do these minor variations come from? Point mutations in DNA, caused by radiation or chemicals, are usually harmful and often edited out of the chromosome by protective mechanisms, so these do not seem a likely explanation for the majority of meaningful evolutionary change. However, duplication of entire genes, caused by some error in chromosomal replication, is a more plausible explanation. In particular, duplication of homeotic genes that guide the development of segmental structures in an embryo is now thought to be a major force in evolution. Simple creatures like nematodes have relatively few homeotic genes, whereas more highly evolved and morphologically diverse creatures like fish, frogs, and humans have larger numbers of homeotic genes.[2] Thus, an understanding of how homeotic proteins regulate gene transcription in a cell has a fundamental place in understanding the evolution of all life.

History has shown that many of our advances in cell biology have originated in accidental observations that were carefully examined. The researchers who found nude, athymic mice or genetically obese mice in their mouse colonies had not maintained these colonies specifically to search for fundamental properties of the immune system or of the hypothalamus. Without these chance observations, however, our knowledge of these topics would be much poorer.

I myself have experienced the influence of accidental observation. I did not become aware of the controversies surrounding oocyte polar cytoplasm or engulfment of cells by megakaryocytes by reading the relatively obscure literature on these subjects. Instead, I found examples of these subjects by accident while preparing histological sections of the ovaries and bone marrow for my students, and wondered what they meant.

The current structuring of scientific research, with its emphasis on a tight scheduling of a specific sequence of experiments all designed to answer yes- or no-questions about a hypothesis, has many strengths. Perhaps, however, this scheme does not adequately allow chance events to influence a path of inquiry. Human beings may not be intelligent enough to even ask the right questions of an enormously complex biological system. Perhaps we should sometimes be content to be led — even haphazardly — to a question by chance instead of being arrogant enough to feel we already know what to ask. Hopefully, both types of approaches will lead to additional fascinating explanations of the behaviors of cells in the future.

References

1. Chan CS, Gertler TS, Surmeier DJ (2009) Calcium homeostasis, selective vulnerability and Parkinson's disease. *Trends Neurosci* 32: 249–256.
2. Lemons D, McGinnis W (2006) Genomic evolution of Hox gene clusters. *Science* 313: 1918–1922.
3. Mayhew CN, Wells JM (2010) Converting human pluripotent stem cells into beta-cells: recent advances and future challenges. *Curr Opin Organ Transplant* 15: 54–60.
4. Taylor R, Lewis M, Powles J (1998) The Australian mortality decline: all-cause mortality 1788–1990. *Aust N Z J Public Health* 22: 27–36.

INDEX